GENT'S

Human Resource Management and Occupational Health and Safety

What springs to mind when thinking about occupational health and safety? Tiresome rules and regulations? A dull, compliance-based aspect of human resource management (HRM)? A litany of sins in the international airline industry, 'burnout' galore in the international call centre industry and failing safety cultures in the international nuclear power industry should shake off some of these commonly held beliefs. Occupational health and safety (OHS) is a complex area that interacts with a broad spectrum of business interests and concerns. However, to date OHS has been confined to the periphery of HRM where its role, influence and importance have, to a large degree, been overlooked. This text sets out to reposition OHS in HRM and business agendas.

In this book Carol Boyd unravels the complex range of factors affecting OHS policy, practice and outcomes. These factors are then placed into context within the international airline, call centre and nuclear power industries, where a wide range of contemporary OHS issues and debates are explored. Of particular interest is the extent to which HRM offers the optimal conditions for 'good practice' in OHS management, which in turn calls for the scrutiny of HRM's underlying philosophy, as well as the impact of HRM principles, policies and practices on health and safety management and OHS outcomes. The author presents a wide range of primary and secondary research where particular importance is attached to separating the rhetoric of HRM, both as a concept and as an approach to people management, from the reality of employees' experiences of contemporary management approaches. In doing so, a number of challenges for policymakers is provided on a range of critical OHS issues.

This book is essential reading for student, practitioner and professional academic audiences who seek a broader understanding of the relationship and interaction between HRM principles, policies and practices and OHS.

Carol Boyd is a lecturer in Human Resource Management at the University of Glasgow. Her main research areas include occupational health and safety and emotional labour and she has had articles published in *Work, Employment and Society*, *Personnel Review* and *New Technology, Work and Employment*.

Routledge Advances in Management and Business Studies

Human Resource Management and Occupational Health and Safety

Carol Boyd

Routledge
Taylor & Francis Group

LONDON AND NEW YORK

Dedicated to Mum, Dad and Brenda

First published 2003
by Routledge
2 Park Square, Milton Park, Abingdon, Oxon, OX14 4RN

Simultaneously published in the USA and Canada
by Routledge
270 Madison Ave, New York NY 10016

Routledge is an imprint of the Taylor & Francis Group

Transferred to Digital Printing 2005

© 2003 Carol Boyd

Typeset in Garamond by Wearset Ltd, Boldon, Tyne and Wear

British Library Cataloguing in Publication Data
A catalogue record for this book is available from the British Library

Library of Congress Cataloging in Publication Data
Boyd, Carol
    Human resource management and occupational health and safety /
    Carol Boyd.
        p. cm.
    Includes bibliographical references and index.
    1. Personnel management–Great Britain. 2. Industrial hygiene–
    Government policy–Great Britain. 3. Industrial safety–
    Government policy–Great Britain. 4. Personnel management.
    5. Industrial hygiene. 6. Industrial safety. I. Title.

    HF5549.2.G7 B69 2003
    353.9'6'0941–dc21

                                                    2002037068

ISBN 0-415-26590-8

Contents

Tables

Preface

What springs to mind when you think about occupational health and safety (OHS)? Tiresome rules and regulations? An under-researched topic? A dull, compliance-based aspect of human resource management (HRM)? A litany of sins in the international airline industry, 'burnout' galore in the international call centre industry and failing safety cultures in the international nuclear power industry should shake off some of these commonly held beliefs.

Occupational health and safety (OHS) is a complex area that interacts widely with a broader spectrum of business interests and concerns. But to date, OHS has been confined to the periphery of HRM, where its role, influence and importance have, to a large degree, been overlooked. This text sets out to reposition OHS in HRM and business agendas.

The text is aimed at student, practitioner and professional academic audiences who seek a broader understanding of the relationship and interaction between HRM principles, policies and practices, as well as a range of contemporary OHS issues and debates. Of particular interest is the extent to which HRM offers the optimal conditions for 'good practice' in OHS management, which calls in turn for the scrutiny of HRM's underlying philosophy, as well as the impact of HRM principles, policies and practices on health and safety management and OHS outcomes.

The book is organised into two main parts, the first of which sets out to unravel the complex range of political, economic and social factors affecting OHS policy, practice and outcomes. These factors combine to provide an explanatory framework for analysing a wide range of primary and secondary research material on OHS policy and practice in the international airline, call centre and nuclear power industries.

I have written this book with a broad audience in mind and with the primary goal of stimulating interest and debate in the subject area. Particular importance is attached to the experiences and perceptions of those employees subject to contemporary management approaches, perhaps at the cost of a more balanced perspective that involves greater contribution from employers. However, I believe this is justifiable on the grounds that the book is directed towards separating the rhetoric of HRM, both as a concept and as an approach to people management, from the reality of employees' perceptions and experiences. In doing so, a range of challenges for policy-makers is provided on a range of critical OHS issues.

Acknowledgements

A number of people and organisations have influenced the content of this book. I have been fortunate enough to work alongside many talented academics, including Michael Quinlan, Peter Bain, Charles Woolfson, Chris Baldry, Phil Taylor, Paul Thompson and the late Harvie Ramsay. Thanks go to all of them for their support, inspiration and guidance. The fundamental role of the Transport and General Workers' Union and the airline cabin crews who contributed to my Ph.D. research is gratefully acknowledged.

Thanks also to my family and friends for their patience and support, as well as to Nick Quinn, Alan Robertson, Peter Bain and Charles Woolfson for their time and suggestions given during proof reading.

Table 2.1 and Table 6.1 are reprinted here by permission of Sage Publications Ltd. Table 2.1 from Quinlan, M., 'The Implications of Labour Market Restructuring in Industrialized Societies for Occupational Health and Safety', in *Economic and Industrial Democracy*, 20, 1999, pp. 427–60, © Sage Publications, 1999. Table 6.1 from Taylor, P. and Bain, P., 'Trade Unions, Workers' Rights and the Frontier of Control in UK Call Centres', in *Economic and Industrial Democracy*, 22, 2001, pp. 39–66, © Sage Publications, 2001.

The author's research on customer violence is reprinted here by permission of Sage Publications and originally appeared in Boyd, C., 'Customer Violence and Employee Health and Safety', in *Work, Employment and Society*, 16: 1, March 2002, pp. 151–69, © BSA Publications Ltd, 2002.

Abbreviations

ACSNI	Advisory Committee on the Safety of Nuclear Installations
AFA	Association of Flight Attendants
APU	Auxiliary Power Unit
ASHRAE	American Society of Heating, Refrigeration and Air Conditioning Engineers
BA	British Airways
BALPA	British Air Line Pilots' Association
BASSA	British Airlines Stewards and Stewardesses Association
BRE	Building Research Establishment
BT	British Telecom
CAA	Civil Aviation Authority
CBA	Cost–Benefit Analysis
CBI	Confederation of British Industry
CIPD	Chartered Institute of Personnel and Development
CUPE	Canadian Union of Public Sector Employees
CWU	Communication Workers Union
DETR	Department of the Environment, Transport and the Regions
DOE	Department of Energy
DOT	Department of Transport
DSE	Display Screen Equipment
DTI	Department of Trade and Industry
DTLR	Department for Transport, Local Government and the Regions
DVT	Deep Vein Thrombosis
ECAC	European Civil Aviation Conference
ECU	Environment Control Unit
EFILWC	European Foundation for the Improvement of Living and Working Conditions
EFQM	European Foundation for Quality Management
EMEA	Europe, Middle East and Asia
ENT	Ear, Nose and Throat
ETCS	European Train Control System
EU	European Union

FAA	Federal Aviation Administration
FARSA	Flight Attendants and Related Services (NZ) Association
HASAWA	Health and Safety at Work Act
HCM	High-commitment Management
HELA	Health and Safety in Local Authorities
HRM	Human Resource Management
HSC	Health and Safety Commission
HSCE	Health and Safety (Consultation with Employees) Regulations
HSE	Health and Safety Executive
HSM	Health and Safety Management
HVAC	Heating, Ventilation and Air Conditioning
IAEA	International Atomic Energy Agency
ICAO	International Civil Aviation Organisation
ICRP	International Commission on Radiological Protection
ICTs	Information and Communication Technologies
IDS	Income Data Services
ILO	International Labour Organisation
INES	International Nuclear Event Scale
IRRT	International Regulatory Review Team
IRS	Incident Reporting System
IRS	Industrial Relations Services
ITF	International Transport Workers' Federation
JAA	Joint Aviation Authorities
JAR	Joint Aviation Requirements
MNCs	Multinational Companies
MOX	Mixed Oxide Fuel
MSDs	Musculoskeletal Disorders
MSF	Manufacturing, Science and Finance (Union)
NEA	Nuclear Energy Agency
NHS	National Health Service
NII	Nuclear Installations Inspectorate
NIOSH	National Institute of Occupational Safety and Health
NRPB	National Radiation Protection Board
NTSB	National Transportation Safety Board
NZ	New Zealand
OD	Organisational Development
OECD	Organisation for Economic Cooperation and Development
OHS	Occupational Health and Safety
OP	Organophosphate
OSHA	Occupational Safety and Health Administration
QRA	Quantitative Risk Assessment
RIAC	Railway Industry Advisory Committee
RIDDOR	Reporting Injuries, Diseases and Dangerous Occurrences Regulations

RMT	Railway, Maritime and Transport (Union)
RNID	Royal National Institute for Deaf People
RSI	Repetitive Strain Injury
SBS	Sick Building Syndrome
SEPs	Safety and Emergency Procedures
SMEs	Small to Medium-sized Enterprises
SOR	Society of Radiographers
SRSC	Safety Representatives and Safety Committees
STA	Science and Technology Agency
TEN	Trans-European high-speed rail Network
TGWU	Transport and General Workers' Union
TOR	Terms of Reference
TPWS	Train Protection Warning System
TQM	Total Quality Management
TUC	Trades Union Congress
UKAEA	UK Atomic Energy Authority
UNSCEAR	United Nations Scientific Committee on the Effects of Atomic Radiation
USA	United States of America
VDU	Visual Display Unit
VOCs	Volatile Organic Compounds
WERS	Workplace Employment Relations Survey
WHO	World Health Organisation
WTD	Working Time Directive

Part I

Mapping the OHS landscape

1 Human resource management and occupational health and safety

Introduction

> The 21st century will be an era of increasingly global competition, and what will distinguish one organisation from another in terms of continuing success ... is optimising the people contribution. Effectively managing the health and welfare of all employees ... will be a major contributor to improving productivity.
>
> (Chartered Institute of Personnel and Development 1995: 4)

Human resource management (HRM) encapsulates a whole range of notions on management theory, style and practice. Perhaps most usefully considered as a generic term that covers the entirety of work organisation, working terms and conditions and representational systems, HRM can be depicted as being concerned with all those activities associated with the management of people in organisations. Storey (1995: 5) defines HRM as 'a distinctive approach to employment management which seeks to achieve competitive advantage through the strategic deployment of a highly committed and capable workforce, using an integrated array of cultural, structural and personnel techniques'. Extensive training and culture management programmes, individualised reward management systems, as well as a range of employee involvement mechanisms, all operate towards achieving enhanced employee contribution. More recently the contributory role of OHS in organisational performance has come into vogue, partly because of alarming upward trends in work-related illness and injury, and the related costs to organisations and governments. Every year about 10 million of the 150 million workers in the European Community are affected by accidents or diseases at work. Direct-compensation costs are estimated at 20 billion ECU per year (Agius 2001). Work-related injuries cost the USA some $125 billion in 1998 – equivalent to nearly three times the combined annual profits of the top five Fortune 500 companies (Economist National Safety Council 1999). For the UK, the cost of work-related injury and illness to employers has been placed at between £3.5 billion and £7.3 billion annually – equivalent to between 4 per cent and 8 per cent of all UK companies' gross trading profits (HSE 1997). This estimation includes the £750 million cost of paying employers' liability insurance to cover compensation to injured workers, the cost of recruiting and training replacements for those forced to give up work and the total cost of property loss. The estimation does not, however, include the costs of social-security payments, NHS costs and the non-financial losses to the victims of accidents and occupational ill-health. When these are incorporated the cost of work-related accidents and illness to society as a whole is estimated at between £10 billion and £15 billion a year – equivalent to between 1.75 per cent and 2.75 per cent of the UK's gross domestic product (HSE 1997). Other estimates have placed the figure as high as £18 billion (DETR/HSC 2000) and, as such, illustrate the haemorrhage both of organisations' profits and of tax-payers' money that follows OHS failures. Despite being alarmingly high, UK official statistics on work-related fatalities, injuries and illness are thought to be grossly underestimated because of the propensity of under-reporting, which makes it difficult to construct a realistic picture of OHS outcomes and related costs. Consequently, the actual cumulative costs of OHS failures are likely to be even more daunting if these figures represent only the 'tip of the iceberg'.

It is surprising to find OHS in such bad condition given that since the 1980s there has been a massive rise in popularity of HRM – an approach to people management that commonly characterises employees as 'the most important asset'. Based on this, an initial expectation might be that OHS will occupy a high-priority position in HRM and management agendas. In addition, HRM is all about making companies more profitable through improved employee contribution. Based on the exponential rise in the costs of employee absence, illness and injury, it would make perfect sense, therefore, to protect and uphold 'good practice' in all aspects of workplace health and safety. But HRM is (allegedly) about more than profit – it is lauded as a 'humane' and 'ethical' approach to people management. However, extracting the value from any asset does not necessarily guarantee a benevolent approach. If workers are treated in as 'rational a way as any other resource' (Storey 1987), is it possible that a trace of malevolence could taint the virtuous HRM? This hints at the critical theorists' accounts of HRM as 'unethical and inhumane' (for example, Hart 1993). These opposing perspectives mean that we cannot be quite so sure that HRM principles, policies and practices guarantee 'good practice' in health and safety management.

While training, selection and culture initiatives have continued to hog the HRM limelight, OHS has to a large degree remained on the periphery. The fact that OHS receives minimal coverage (or none at all) in key HRM texts and journals suggests that commentators have been slow in recognising the interaction between a range of people-management policies and work-place health and safety. One exception is Bach (1994), who argues that HRM and health and safety are intimately connected by four themes: cost effectiveness, commitment, quality and strategic integration. In his view, cost efficiency may be undermined by the cost of accidents, in terms of the costs of absence and fines resulting from Health and Safety Executive (HSE) action, while the management of quality and safety may affect the ability of companies to produce quality output consistently. Further confirmation of this is found in literature from the Chartered Institute of Personnel and Development (CIPD) – the professional body of human resource/personnel specialists – where OHS is identified as a critical factor in making organisations more profitable, productive and cost-effective. From government, the same message is echoed:

> the health and well-being of employees is a key factor in the success of any business or organisation. Recognising the value of a healthy workplace will ensure that staff are 'healthier, happier and here'. Placing these issues at the centre of an organisation's concerns will help ensure its continuing effectiveness.
>
> (*The Healthy Workplace Initiative*, UK Department of Health,
> April 1999)

However, the ideological stance that good health and safety management contributes to organisational success is compromised by the common reality that financial imperatives often undermine good health and safety practice, explaining in part why all too often organisations fail to practise what they preach, as evidenced by an array of international industrial tragedies (for example, Piper Alpha, Tokaimura nuclear accident, Southall rail crash).

While our expectations of health and safety management under a range HRM principles, policies and practices may be fuelled by the rhetoric of its proponents, the economic logic for 'good' OHS practice along with the expressed policy intentions of policymakers, the different emphases given to HRM are also worth noting. These are contained in the 'hard–soft' dichotomy. In simple terms, these emphases occupy polar ends of the HRM spectrum, with 'resource to be invested in' or 'soft HRM' at one end and 'cost to be minimised' or 'hard HRM' at the other. The 'hard' model reflects a 'utilitarian instrumentalism', where the overriding characteristic is the focus on the 'quantitative, calculative, and business strategic aspects of managing the headcount resource in as "rational" a way as for any other economic factor' (Storey 1987: 6). Short-termism and cost prioritisation are two of the potential dangers (and often realities) of such an approach to people management and OHS. Alternatively, the 'soft' developmental approach, while still emphasising the importance of integrating human resource policies with wider business objectives, stresses the development of employee commitment via 'communication, motivation and leadership' (Storey 1987: 6). However, it does not usually come down to a case of either/or. A pick-and-mix approach combining both 'hard' and 'soft' aspects of HRM is visible in a number of major organisations, including Hewlett-Packard,

Marks and Spencer and British Airways (BA). Keenoy (1997) describes how BA has taken a 'hard' approach to 'headcount', while at the same time implemented a wide-ranging programme of 'soft' HRM (see, for example, Truss *et al.* 1997). As such, 'HRM' remains a 'slippery concept', able to mutate and conform to organisational needs in rapidly changing economic and political environments where a cost variable status may be all too easily assigned to OHS.

The debates about HRM are to date extensive, but few, if any, have explored this approach to management from an OHS perspective. As such, OHS provides an ideal prism through which we can explore one of the most popular and debated approaches to people management in recent decades. This text sets out to provide a unique litmus test for the faith many commentators have placed in HRM principles, policies and practices. In doing so, we consider some of the fundamental principles within HRM, such as consensus in the employment relationship, employee involvement and some of the structures that attempt to facilitate 'communication, trust and openness' between employers and the employed. In addition to this, the primary goal of improving employee contribution through an 'array of strategically integrated personnel techniques' (Storey 1995), directs our attention to how HRM policies and practices impact upon work organisation. When these components are set in an organisational context, particularly within current political and economic climates, we may be more inclined to question whether HRM is even capable of offering the conditions conducive to good policy and practice in OHS.

HRM and OHS: was it good for you?

Throughout the 1980s and 1990s HRM gained massive popularity and attracted considerable attention from both practitioners and academics (Hill 1991; Marginson *et al.* 1993; Millward *et al.* 1992). For some commentators, the ascendancy of HRM can be understood partly on the basis of 'right time, right place' (e.g. Keenoy and Anthony 1992; Legge 1995). As Hyman (1995: 28) notes, from 1979 conservative industrial relations since the industrial revolution', where a myriad of potent factors led to the transformation of the structures and processes immanent in the employment relationship. Bubbling in the cauldron of change were economic recession, globalised markets, rapid technological change, culled employment legislation, the feminisation of the workforce and the proliferation of Japanese and US multinational companies across the UK – all of which created fertile conditions for the propagation of HRM principles, policies and practices. Momentum continued to build as proselytes preached to the masses on the almost magical abilities of HRM to transform employee and organisational perfomance, while the subject area went on the gain academic credibility following the proliferation of international journals and texts covering the area. This flurry of activity and attention was instrumental in repositioning the personnel practitioner's role from the downtrodden 'middleman' into a high-flying HR executive charged with strategic responsibilities.

For some commentators, the HRM decades have provided little evidence of a new 'ethical' approach to people management, with UK organisations 'locked into a vicious circle of low pay, low skill and low productivity' (Sisson 1994: 42), and the regrouping of *laissez-faire*, short-term, cost-reduction employer attitudes (Ackers 2001) combining to offer a gloomy presage for OHS. However, amidst these tumultuous conditions, OHS may have enjoyed the sheltered haven provided by the commonly perceived 'greater degree of shared interest' between employers and employees in relation to health and safety. Indeed, with a proposed 'all new' consensual approach to employment relations, many of the ruffled feathers in the workplace might also be smoothed. Hopes were raised, and as Ackers (2001: 391) explains:

> The sleek new HRM model is boldly contrasted with the 'bad old days' of personnel past, much as a born-again Christian celebrates his new creation by darkening his own past. Before, labour was a cost to be controlled; now, a resource to be nurtured. Before, personnel was a routine administrative activity; now, a strategic champion of people management for heightened business performance. Before, industrial relations was adversarial and arm's length; now, founded on consensus and employee consent.

Quite a feat, especially when we consider the extent of the transformation in employee and management attitudes, especially over historically conflictual issues. According to some commentators, OHS is not, and has never been, a matter of general consensus between employers and the employed (Carson 1985, 1989; Carson and Henenberg 1988; Nichols 1997). These commentators might direct our attention to the evidence that identifies health and safety as one of the most common workplace grievances and reasons for ballots for industrial action (TUC 1999a, 2000, 2001a). They might also point to the damaging impact of employer prerogative on workplace health and safety outcomes as well as to the correlation between trade union influence and levels of employment protection. For example, evidence suggests that higher injury and illness rates occur in workplaces where management alone decide on health and safety issues (Millward *et al.* 1992; Reilly, Paci and Holl 1995; Walters 1996; James and Walters 1997; Nichols 1986; Litwin 2000).

Since managerial prerogative is a cornerstone of HRM principles and practices, the above evidence might lead us to lose confidence in the ability of HRM to adequately provide for 'good practice' in OHS. However, the emphasis that HRM places on communication and employee involvement could overcome some of these concerns. For example, team briefings, problem-solving groups and newsletters are all highly popular communication techniques in the UK (Cully *et al.* 1999). In addition, employers are under a legal duty to consult with their employees, either directly or through a representative, on health and safety issues. However, some critics have argued that these consultative forums offer little in the way of meaningful employee participation, limiting in turn workers' ability to have their interests heard and addressed. This is of particular concern in relation to workplace health and safety, given the range of evidence that suggests that where employee representatives have a meaningful participative role in OHS, fewer accidents and injuries occur (for example, Millward *et al.* 1992; Reilly, Paci and Holl 1995; Walters 1996; James and Walters 1997; Nichols 1986; Litwin 2000).

It appears then that HRM principles and practices are struggling to deliver promises of consensus, trust, meaningful employee involvement and open communication in the workplace. Can we expect similar difficulties in the area of employee performance? It appears that this is one area where the short-term returns on HRM policies and practices have been impressive. The implementation of reward, appraisal and training strategies has, by some accounts, led to remarkable improvements in both individual employees' and organisations' performance (for example, read any CIPD material). Bathing in the spotlight of individual accountability, some employees have flourished while others have undeniably withered. According to one individual working under HRM-related principles, policies and practices in a manufacturing plant: 'Don't fall ill, don't grow old, and above all, don't get tired' (Delbridge and Turnbull 1992: 69).

This points to a possible malevolent streak in HRM – how far will HRM policies and practices go to achieve (short-term) enhanced employee performance? Short-term benefits are flagged up because of the evidence pointing to the longer-term outcomes of these practices on employee health. Absence and turnover costs, never mind employee compensation for work-related injury, are, after all, potentially huge drains on profits. However, who said HRM was about the long term? Evidence from the international call centre industry points to a 'sacrificial HR strategy' whereby stringent recruitment and selection techniques cherry-pick the most vibrant and enthusiastic recruits who are subsequently bled dry by an unrelenting intensive work regime. As one of the managers explains: 'We don't want people to stay past 18 months. By that stage they are burnt out and are no good' (Wallace, Eagleson and Waldersee 2000: 178).

The clear interaction between work organisation and employee health flags up an important gap in the literature. While there is voluminous evidence relating to the outcomes of culture management, training and customer service programmes, it appears that a lesser emphasis is given to the interaction between these people management policies and employee health. This of course might expose a side of HRM policies and practices that employers, consultants and policymakers fail to, or choose not to, see.

In forthcoming chapters we explore possible linkages between these policies and employee health and safety. While doing so might be useful for practitioners and employee representatives in presenting an economic or 'business' case for adaptations to working practices for example, it is important not to overlook a major caveat. In reinforcing OHS as a cost variable, nothing is done to address the fundamental weaknesses in approaches to health and safety management in the UK and elsewhere. These weaknesses relate to the political and economic environments in which HRM principles and OHS are immersed. They also suggest that in HRM, we were backing a loser from the start.

Supporting logic for this position is provided by Ackers (2001: 382), who argues that HRM's philosophical underpinnings are grounded in right-wing ethical thinking, which stress the utilitarian benefits of free market capitalism, duty to the shareholder, justice as the entitlement to own and freely dispose of property, and a paternalist version of enlightened self-interest that renders state (and trade union) intervention unnecessary. According to this view, HRM is founded on and driven by business (profit) rationalisations, where the goal is to produce valued *economic* assets. In this worldview, human rights are entirely conditional on business convenience and what makes business successful. According to Ackers (2001: 393), the price is personnel's professional soul and their professional autonomy and integrity in their accepted role as the handmaidens of private capital. In unrelenting condemnation of HRM, Ackers (2001: 395) concludes that the 'abdication of any autonomous, ethical role for management exposes HRM as a public relations façade and rationalisation for what business already does out of short-term economic self-interest'. Based on this critique, the prospects for OHS appear bleak. Firmly positioned as a cost variable within voluntaristic regulatory frameworks that espouse minimal state intervention in OHS as much as any other area of the employment relationship, and driven by economic logic, it would follow that employers will only invest in OHS if this worked out to be cheaper than the costs of not doing so (for example, fines, compensation, turnover costs). This brings us back to the question presented earlier, 'is HRM capable of providing the optimal conditions for "good practice" in OHS management?'

To establish this, three main issues need to be addressed. First, to what extent have HRM principles, policies and practices created consensus in the employment relationship and, by implication, workplace health and safety management? Second, to what extent do HRM-related employee involvement mechanisms adequately represent and reconcile employee interests and concerns over OHS? Third, to what extent do HRM policies and practices lead to work intensification in terms of higher demands being placed on both emotional and physical labour? These questions relate directly to the first four chapters of the book, where in Chapter Two we explore the basis of Ackers' and other commentators' criticisms of HRM's underlying philosophy and principles in some detail, while also illustrating how these assumptions feed into the UK regulatory framework for OHS. In Chapter Three we focus on consensus in the employment relationship, with particular regard to OHS, and the processes and structures that are available for the resolution of any conflicts of interest. Chapter Four then goes on to highlight a range of workplace factors in OHS, with a particular emphasis on work organisation and the physical and emotional dimensions of labour. Together, these first four chapters unravel many of the factors involved in the international (mis)management of OHS. These factors are then placed into context in Chapters Five, Six and Seven, where in-depth case studies of HRM and OHS in practice in the international airline, call centre and nuclear power industries illustrate the complex and synergistic relationship between HRM policies and practices, and workplace health and safety. Chapter Eight provides some observations, conclusions and recommendations as to how OHS might be taken forward in the future.

2 The regulatory politics of OHS

In this chapter we focus on the range of political, economic and social forces that influence policymakers such as the pressures created by globalisation, deregulation and privatisation. In a number of industrialised countries, political ideologies favour right-wing neo-liberal economic theory, affecting the form and tone of policies, such as those governing the employment relationship. The knock-on effect on OHS is apparent in the case of labour market restructuring, which has been intimately connected to increasing job insecurity, and the related propensity to suffer work-related illnesses. This range of macro-level factors provide the backdrop to OHS and to a range of HRM policies and practices, making an understanding of their interaction with HRM and OHS essential.

In the last chapter, our review of HRM cast doubt upon any presumption that 'good' OHS practice is assured by an HRM approach to people management. It was proposed that HRM's basic congruence with prevailing economic theory means that it is driven by individualism and business (profit) rationalisations, and based on this occupational health and safety interventions may be entirely conditional on business-led decisions. At first sight, this assumption does not necessarily lead to negative consequences for OHS. It could be argued that business rationality would base itself on the relative costs of 'good' OHS practice and OHS failures. For example, good practice may protect the organisation from increased insurance premiums, a tarnished reputation, civil-damage claims, fines and the costs associated with lower productivity, employee absence and turnover – all of course based on the assumption that the relative costs of 'good' OHS and related interventions outweigh the costs or losses if these were withheld. Such logic is embodied in a 'self-interest model' – mutated from economic theory – which holds that employers willingly and consciously follow safe practices, and implement and adhere to preventative strategies in response to market pressures, the existence of a constraining legal framework as well as actively avoiding sanctions and costs associated with OHS failures (Viscusi 1979; Dorman 1988, 1996; Moore 1991).

Critics might argue that markets do nothing to protect workers from unsafe working conditions and instead act to exacerbate existing inequalities in the workplace. At the same time the often flimsy restraints provided by legal frameworks suggest that governments bend towards business interests when not actually captured by them. Such biases are arguably reflected in the minimalist nature of OHS legislative frameworks in the UK and USA, the declining resources of regulatory agencies (for example, the HSE in the UK and OSHA in the USA) and the often derisory penalties for OHS violations in these countries. In addition to these indicators are the ubiquitous cost benefit exercises in OHS. As indicated earlier, where the cost of providing OHS interventions is greater than the cost of withholding them, free market logic would condone doing just that.

While keeping sight of cost efficiency in OHS interventions is important, it may be difficult for some organisations not to let this be their guiding mantra. For example, in 1978 the car company Ford decided not to undertake the necessary technical modifications to its Pinto model that would reduce the risks of fuel-tank explosions. This decision was based on a cost benefit analysis that calculated financial savings of US$87.5 million, despite 180 predicted fatalities. This example suggests quite crudely that even leading corporate concerns can knowingly permit safety to be compromised for profit. While shocking, this behaviour is not particularly unusual. A number of public inquiries into industrial accidents have identified cost-cutting exercises and wilful ignorance on the part of employers and managers have been identified as a major contributory factor (for example, Piper Alpha, the Clapham rail crash, the King's Cross fire, the loss of the *Herald of Free Enterprise* and the Valujet disaster[1]). The philosophical underpinnings of a cost benefit analysis approach to OHS can be located in neo-classical economics – an area worth exploring if we are to increase our understanding of past and present approaches to OHS.

Voluntarism and free market economics

At the centre of neo-classical economics is the belief that the pursuit of self-interest by capitalists, workers and consumers leads to the most efficient utilisation of an economy's resources. According to this perspective, freedom from state interference in the labour market allows workers to deploy their skills to their best advantage in the labour market, while employers are able to respond to the demands of consumers in terms of goods and services. The variables of pay (and other working conditions such as safety), productivity, goods and services will be 'regulated' by the subsequently created market forces that in turn reflect the rational choices of individuals. Such are the arguments that have driven forward political, economic and social reform in industrialising countries since the eighteenth century. However, there are a number of unstable assumptions in this perspective.

The first assumption purports that levels of risk and industrial illness/injury are not too high because of the inhibiting effect of the greater cost of compensating workers for taking on dangerous work. This supposedly provides an incentive to employers to make working conditions safer – that is if the cost of doing so is less than the corresponding wage differential (Dorman 1996: 27). However, Dorman (1996: 137) goes on to argue that there is little evidence to show that firms actively take wage compensation into account when deciding on working conditions. In addition, those workers on lower pay are often offered more hazardous working conditions. Based on this, Dorman (1996) challenges Smithian notions of equity and efficiency as well as the principles of wage-compensation theory. In addition, he argues that that 'self-interest' for workers is often set against a background of finding employment within labour markets that have pools of involuntary unemployment, while for those in work, the threat of dismissal or redundancy looms overhead. Consequently, the extent of employees' 'free choice' will be somewhat more limited than that of employers, particularly when it comes down to 'accepting' the presence of risks in the workplace.

Related to this is the belief that intervention from the state (for example, providing employment protection rights) in the private contractual relationship between employers and employees constitutes disruptive interference and will in turn lead to economic inefficiency. The bottom line appears to be that 'free choice' allows employees to conduct their 'risk analysis', and in accepting the job they have also accepted the concomitant risks. If the state were to intervene with remedial regulation that effectively removed dangerous jobs, then this would only serve to disadvantage workers through removing 'free choice' and the associated extra dividends. However, employees may not always have all of the information about the range and extent of risks involved in a job, nor may they be in a position to refuse such work, because of financial circumstances for example. Dorman (1996: 206) cites one study that found when employees were provided with the 'right to refuse' dangerous work, 93 per cent of refusals came from unionised workers. However, he cautions, two-thirds of all (US) workers are not unionised and it is often these workers who face the most extreme risks. In addition, the assumption that dangerous work carries more financial or other rewards is not upheld by empirical research or injury statistics (see Dorman 1996 and Nichols 1997 for a full discussion).

A further assertion of economic theory is that safety is another commodity bought and sold in the marketplace where its price reflects the value as derived from employer and employee preferences and where the relative costs of provision and costs of withholding safety measures are weighted in some form of cost–benefit analysis (CBA). However, CBAs are rejected by the Organisation for Economic Co-operation and Development (OECD) on the grounds that these 'lead to socially unacceptable levels of occupational injury, disease and death' (OECD 1986) – which is the same conclusion drawn by policymakers in the early nineteenth century following the *laissez-faire* experiment of that period.

Determined not to learn anything from history, there was something like a revival in the late twentieth century, which, as Dorman (1996: 31) is quick to point out, reflects shifting balances of political power rather than the actual virtues of the approach. Across many of the developed countries, free-market economics have been used to justify the legislative weakening of trade unions, the repeal of protective employment legislation and a more generalised resistance to statutes designed to improve the conditions of work. For example, the 1980s saw step-by-step legislative attacks on UK trade union organisations and dilution of protective employment legislation (see, for example, Bain 1997), while in the 1990s New Labour made only half-hearted attempts to redress what was a fundamental shift in the balance of workplace power towards employers.

And just as government and employer attitudes and policies seem to mirror history, so too do trends in workplace death, injury and disease. In early Victorian times, the horrors of unregulated capital in the form of large-scale death, injury and industrial disease were to stimulate wide-ranging social criticism from leading commentators including Engels, Mayhew and Booth. However, by the end of the nineteenth century it became acceptable to hold that private market over regulation would inevitably lead to unacceptable levels of industrial death, injury and disease (Moore 1991: 7) despite the introduction of the first Factory Acts. More recent figures on workplace injury and illness support the argument that twentieth-century free-market economics are as uneconomic and inefficient as in earlier periods. For example, the UK Workplace Employment Relations Survey (WERS) (1998) reports that in the previous year, at least one employee suffered a serious injury in 61 per cent of workplaces (Cully *et al.* 1999). In addition, Health and Safety Commission (HSC) figures for the UK report a 1 per cent rise in reported injuries in 1999/2000 compared to the previous year, as well as a gradual growth in the number of reports concerning work-related illnesses such as pneumonconiosis, occupational asthma and dermatitis, and musculoskeletal disorders (HSC 2000a). Statistics for the following year show that the number of workplace fatalities increased by 34 per cent (295 up from 220 the previous year), with an estimated 7,800 *new* cases of work-related musculoskeletal disorders and some 6,600 *new* cases of work-related stress and mental-health problems seen by specialist physicians in that period (HSE 2001a). As a result of extensive under-reporting, estimated to be as high as 40 per cent, the official figures are regarded as representing only the tip of the iceberg (HSE 2001a). In explaining these figures and trends, some attention must be given to the regulatory framework.

The regulatory framework

Following the failed *laissez-faire* experiment in the nineteenth century, a raft of regulation was created to protect the health and safety of UK workers. Regulations mushroomed and multiplied, driven forward by the increasing power of trade unions and sympathetic governments. However, by the

1970s considerable reform to the complicated array of workplace regulation was underway. These were the conditions underpinning the Robens Committee review of safety at work (Robens 1972), which culminated in the 1974 Health and Safety at Work Act (HASAWA). This Act not only changed the approach to industrial safety in the UK but its influence extended to other countries, particularly Australia.[2] The committee's main aim was to replace the inherited mass of detailed, prescriptive regulation with a more rational, broad, goal-based regulatory framework (Robens 1972). The Robens Committee proposed a single comprehensive unifying and enabling piece of legislation that would lay out the basic duties of both the employer and the employee. Safety was to become primarily a matter of 'self-regulation' rather than prohibitory, external regulatory control (Robens 1972). However, the report was, at that time, criticised on a number of grounds, and later by Woolfson and Beck (1995) and James (1992). According to Nichols and Armstrong (1973: 30):

> The Robens report was largely written by administrators, the kind of people for whom, maybe, the thought comes hard that the real safety and health problem is to protect workers against the inherent 'unnatural' excesses of a society dominated by the market; a society in which some men are paid to squeeze as much production as possible out of others . . . Most of all, they (the authors of the report) never realized that in a society deeply divided between those who control and those who are controlled, goodwill, however much of it exists, is simply not enough . . . people who do the producing must have the power to ensure that their safety is put first.

Nonetheless, the HASAWA (1974) was adopted and to date, it has provided the basic framework for the regulation of OHS in the UK and a model for many other jurisdictions.

The HASAWA is characterised by a 'voluntaristic' approach whereby the state retreats from regulatory intervention and supports self-regulation by industry. The rationale for self-regulation is explained by the Robens Report (1972: para. 114):

> A principal theme of this report is the need for greater acceptance of shared responsibility and more reliance on self-inspection and self-regulation and less on state regulation. This calls for a greater degree of real participation in the process of decision-making at all levels.

However, 'real participation' was not secured via the HASAWA, but through a later legislative victory won by the trade unions following a long campaign. The Safety Representatives and Safety Committees (SRSC) Regulations (1977) provided a strong platform for UK trade unions to assert their rights to a safe and healthy workplace. The importance of meaningful employee participation in workplace safety is well illustrated by the fact that the relative strength of capital (employers) and labour (trade unions) corre- lates with the apparent initial success of the Act in reducing workplace acci- dents and injuries during the late 1970s, while the subsequent rise in these statistics during the first half of the 1980s correlates with the ascendance of Thatcherite union-bashing policies (Nichols 1997). One interpretation of these findings is that when their role is legitimised by legislated rights that provide an effective counterbalance to the might of capital, trade union involvement will lead to an improvement in workplace safety. This view is supported by Reilly, Paci and Holl's (1995) findings, despite finding a posit- ive association between union presence and higher injury rates. Nichols (1997: 151) explains this curiosity on the grounds that 'history suggests that it is not trade unions that call forth unsafe work conditions but rather unsafe work conditions that call forth trade unions'. Indeed, a range of US studies show that unions are more common in high-risk industries (Leigh 1982; Worrall and Butler 1983; Hirsch and Berger 1984, quoted in Nichols 1997).

As we will see in the next chapter, a range of structural and sectoral shifts have served to underline the 'passed sell-by date' of HASAWA, such as the shift away from large, single-entity employers to multiple small enterprises. The UK, like many other industrialised countries, has experienced a land- slide shift towards smaller enterprises with a large proportion of (non- unionised) part-time workers. According to recent research (Cully *et al.* 1999; Millward, Bryson and Forth 2000) the typical UK workplace size is just over 100 employees, with small to medium-sized enterprises (SMEs) accounting for 99 per cent of all businesses in the UK (DTI 1998). Small businesses[3] account for over 40 per cent of total employment in many indus- trialised countries (Burgess 1992; Wiatrowski 1994; Witmer 1997). Con- sequently, a large proportion of the working population may find themselves subject to nebulous OHS rights and protection.

As more of an ideal vision than an actual reality, Robens' notion of self- regulation embodies a sterile image of employers and employees co- operating in workplace health and safety – an image that fits well with the 'happy-clappy' land of consensual HRM. In this land, power differentials or opposing interests do not exist, nor do competing production and economic pressures. As Nichols (1997) has argued, Robens' underestimation of the profit imperatives held by capital and the assumptions about consensus and equality in the employment relationship (two features not normally associ- ated with it) serve to underline the shaky foundations of OHS in the UK – foundations that were soon to be disturbed by seismic political tremors.

The changing legislative framework

From 1979 in the UK, the new Conservative government embarked upon the wholesale deregulation and the privatisation and liberalisation of indus- try. In October 1992, the UK government launched the Deregulation Task

Force, which conducted a review of an initial 400 pieces of regulation. The fruits of the exercise are shown in the 1997/8 Health and Safety Commission (HSC) annual report, where it is noted that fifty-three sets of regulations and Acts had been removed as part of the ongoing programme of legislative reform. Bain (1997: 187) cogently sums up the reform campaign:

> The deregulatory bandwagon, which has been rampaging along the highways and byways of US and British industry for several years now, shows few signs of stopping of its own volition. The fact that health and safety legislation has been assailed during its unruly progress is not because the vehicle accidentally strayed from its designated path, rather that these regulations were targeted as part of the overall 'search and destroy' mission.

Part of this 'search and destroy' mission has been the restrictions placed on state enforcement agencies. In the 1980s, the government's goal of limiting the cost impact of new health and safety legislation on business was supported by its instruction to the HSC to consider the economic implications of any proposed new regulations. In addition, during the 1990s the government pressurised the agency to adopt a less rigorous approach to enforcing the legislation. Under a new code for enforcers, inspectors were advised to adopt a more sympathetic attitude to business problems and to adopt a 'softly softly' approach to the enforcement of health and safety legislation (for a full discussion see Burrows and Woolfson 2000). More recently, the HSE/C has been subject to requests from the government to take on a greater advice-giving role, rather than involvement in policy formation (IRS, April 1999). Effectively, the HSC has been under continual attack not only in its underlying philosophy and ethics, but also in financial terms with swingeing cuts to budgets during the 1990s. In 1993, the status and muscle of the agency atrophied when the enforcement operations were subjected to 'market testing' to see if the private sector could provide services more cheaply, and the job of chairing the HSC was relegated to part-time status by the government (Bain 1997: 183).

The underlying ethos of maintaining minimal state intervention and allowing employers relative freedom to prioritise the imperatives of profit and cost-cutting is particularly well illustrated by the UK's reactions and responses to European influence. Without doubt, many recent improvements to OHS regulation in the UK do not arise from domestic pressure but emanate from Europe. European Directives such as regulations on minimum wage, working time and parental leave have cumulatively led to a degree of improvement in workers' employment rights and protection across Europe. However, consecutive UK governments have sought to weaken and dilute European Directives with a range of exclusions, provisos and opt-outs. One example is the EU Working Time Directive, which produced the UK Working Time Regulations (WTR) (1998). The thrust of the WTR was to place an upper limit on employee working hours (an average of 48 hours per week) in response to health and safety concerns about the 'long-hours culture' and the absence of rest breaks in some industries. Eurostat (1995) figures revealed that full-time UK employees worked the highest average number of hours (43.3 hours compared to a European Union (EU) average of 40.3 hours) in EU member states, with almost one-fifth of full-time employees in the UK working more than 48 hours per week – nearly treble the EU average and a little over Ireland, the next highest contender.

UK figures for 1998 show that one in eight full-time employees routinely worked more than 48 hours per week, with the highest concentration amongst managers and those employed in construction (Cully *et al.* 1999). The propensity to work long hours does not appear to have been dented by the WTR. Goss and Adam-Smith (2001) report from a study of 416 private sector companies that only 13 per cent of companies had responded to the WTR by reducing the hours of work. The most common response was individual agreements (81 per cent of employers surveyed) – usually containing some form of 'opt-out' from the WTR restrictions on hours of work. The ubiquity of 'opt-out' arrangements within individual agreements over working hours is further illustrated by a recent DTI report, which found that the large majority of companies surveyed were using individual opt-outs (DTI 2001). This situation reflects the positioning of occupational health and safety on government and employer agendas. Despite the wealth of evidence that links long working hours to ill-health and confirmation from the European Court of Justice that working time is an OHS issue rather than just a matter of general industrial relations (which came in response to a challenge by John Major's Conservative government), defiant policymakers have sided with business concerns over flexibility and efficiency. This is evidenced by the diluted form of the regulations, which came complete with a range of derogations and bare-faced guidance on how to exploit the loopholes. Despite the good intentions of European policymakers and the superficial attractiveness of the regulations, it appears that the UK government has successfully reshaped the nature, form and implementation of the UK Working Time Regulations to fit with the needs of capital. In the words of Goss and Adam-Smith (2001: 205), 'the apparent protections offered by greater jurisdiction [of the European Union] may turn out to be janus-faced'. As Arrowsmith and Sisson (2001) note:

> So far ... the most immediate effect [of the WTR] has been to encourage employers to seek 'flexibilities' and 'opt-outs' to minimise their impact by agreement with employees.

Interestingly, Goss and Adam-Smith (2001: 206) report that individual agreements over working hours were most common in companies with trade union representation – a finding they explain on the basis that such compromises are made under conditions of ever-present job insecurity. As such, these agreements may constitute a pragmatic response by a growing proportion of the UK workforce who perceive long working hours as part of the price that has to be paid for relative security. The authors further note the uneasy position of trade unions, which may not approve of such agreements in principle but may have little choice other than to acquiesce. We can link this finding back to assumptions about employees' freedom to choose and equity in the employment relationship, which bear little likeness to the reality of work and employment relations.

We now move away from legislation and regulatory frameworks towards the broader impact of free-market economics taking us into the fast-track environment of globalisation, liberalisation and privatisation. This provides an opportunity to consider the relative strengths and weaknesses of neo-liberal economic theory and free-market principles in fast-changing markets.

Privatisation and liberalisation of industry

The adoption of free-market economic policies in industrialised countries has led to the increasing globalisation and liberalisation of markets and to the extensive privatisation of state-owned enterprises. Recent years have seen the spread and growth of multinational companies and increasing competitive pressures within (and between) countries, along with the privatisation of many publicly owned industries (for example in the UK, British Gas, British Telecom, British Airways) as well as liberalisation in particular sectors (for example, the docks and the financial sector).

Since the early 1980s, some forty former UK state-owned businesses have been privatised, a process that affected more than 600,000 workers (*Guardian*, 22 November 2000). As the newly cost-conscious companies responded to opportunities to boost productivity and profits through restructuring, many thousands of jobs were lost. The legacy of privatisation is observed in the railway industry. As privatised rail companies restructured their operations, skilled employees with long experience were lost in the industry as subcontracting proliferated with tragic consequences.[4] Nonetheless, New Labour appears determined to privatise London Underground, with concerns over the proposed fragmentation of safety systems becoming a point of heated contention between unions and government, suggesting in turn that few lessons have been learnt. The dangers of placing our faith in market forces are well illustrated by a number of international examples.

Placing our faith in market forces

The guiding assumptions made by neo-liberal economists is that market forces act as a decisive rationalising force in ridding industry of incompetent management and poorly performing organisations. A secondary argument is that it is in the interests of business in general to ensure adequate safety and health for business-led reasons if no other:

Some enforcers seem to believe that business people seek to cut corners
and ignore safety. Businesses are impelled to place safety at the top of
their priorities; it is in their own best interests to do so. A restaurant
where the customers fall ill will soon lack customers; an airline or bus
operator with a poor safety record will find passengers reluctant to use
it. Business people generally apply themselves with great dedication to
ensure that their businesses are safe for employees, customers and neigh-
bours.

(Department of Trade and Industry 1994: 1)

The first point to note is the notion that businesses, in theory at least, are at
prior liberty to harm customers' safety and/or health. The restaurant customers had to suffer food poisoning before the restaurant acquired a bad
reputation, while the airline or bus company likewise had to kill or maim
some employees or passengers before business suffered. The second point is
the suggestion that unscrupulous employers will cease to trade because
society, through market forces, reacts to bad practice. Yet following various
industrial accidents (for example, the Clapham rail crash, the King's Cross
fire, the Piper Alpha disaster, the loss of the *Herald of Free Enterprise* and the
Valujet disaster[5]), the employers in question continued to operate as functioning business concerns. Valujet is still in business, albeit trading under a
different name (Transair), while Occidental Petroleum, operator of the Piper
Alpha oil platform, continues as a global oil player. The rail operators,
however, have had their wings severely clipped, proving that societal tolerance of continued corporate safety failures has its limits – even in a climate
hostile to state regulatory intervention. A closer look at developments in the
UK railway industry provides ample evidence of the inherent weaknesses in
free market logic.

Formally nationalised in 1947, the UK railway industry was progressively
privatised under the Railways Act 1993 between 1994 and 1996. Since then
a number of serious rail crashes have occurred at Southall, Ladbroke Grove
and Hatfield with a total of forty-two lives lost. Subsequent public inquiries
listed a range of recommendations relating to safety improvements – all of
which required massive financial investment. The preceding Conservative
government had decided that although a train-protection warning system
(which overrides the driver's control and can stop the train in the event of an
impending collision) could save fifty-two fatalities over 20 years, the cost of
£14–15 million per life saved was too expensive (*Health and Safety Bulletin*
262, October 1997). A cheaper, downgraded version was subsequently
installed, the efficacy of which is tragically observed. Although serious accidents occurred while the railways were nationalised (for example at Clapham
Junction in 1988), privatisation has arguably aggravated the problem of
inadequate levels of investment in safety. According to Gerald Corbett, Railtrack's then chief executive, the rail network was 'like a pressure cooker
waiting to explode' (*Guardian*, 11 November 2001).

Following the Southall Inquiry, the government introduced a new Transport Bill, which has been enacted as the Transport Act 2000. From this, The Strategic Rail Authority was formed. This agency has new power to control and direct investment into the rail industry, amounting to £49 billion over 10 years to cover the installation of upgraded train-protection warning systems. While entirely welcome, the reactive nature of this intervention should not go unnoticed. The move is also likely to be a delayed response to the requirements of the 1996 European Directive on interoperability for the Trans-European high-speed rail network (TEN), which covers four UK lines (Great Western Main Line, West Coast Main Line, East Coast Main Line and the Channel Tunnel Rail Link).

Presently, the train-protection warning systems (TPWS) on these railway lines do not comply with these regulations. Despite information being available since 1991 on the content of the Directive and the publication of a first draft in 1994, the downgraded TPWS was fitted in the UK as a simple, cheap and quickly available means of providing train protection. According to one commentator, the UK is 'just about the only country left in the world that runs high-speed trains and mixed traffic without the benefit of modern automatic train protection systems that prevent the driver passing signals at danger' (HSE 2001b). In complying with the European Directive, the European Train Control System (ETCS) or equivalent will (eventually) be fitted.[6] This system makes use of new radio communications technology to create an entirely new command and control system for the railway (HSE 2001b). Effectively, the UK rail companies have had to be forced by (supra)state intervention to provide adequate safety standards.

All of this goes some way to highlight the failure of voluntaristic approaches to producing high-quality (or ethical) OHS standards. Nonetheless, the UK government displays similarly stubborn and quixotic views of OHS regulatory framework in the UK. According to John Prescott, the Deputy Prime Minister:

> The [HSAW] Act provides a framework for good, effective regulation that has transformed Britain's workplaces. We can see the results – the number of deaths at work today is a quarter of the 1971 level.
> (Foreword in *Revitalising Health and Safety: Strategy Statement*, DETR/HSC, June 2000)

This boast needs to be challenged on the basis that industries with high fatality rates have declined significantly since the 1960s and 1970s. Indeed, mining, docks and shipbuilding are portrayed as male-dominated 'dinosaur' industries in the UK today. By contrast, the burgeoning service sector is represented in the main by women and young workers. In addition, this sector is home to a range of new production technologies. While fatalities in this sector and other sectors are relatively uncommon, *injury and illness* rates are far in excess of those recorded in the 1970s. The 'transformation' in

British workplaces relates, therefore, to the reduced risk of being killed at work and the increased risk of suffering a work-related injury or illness (for example, RSI or a stress-related illness).

Many of the structural and sectoral shifts occurring in the UK are mirrored in other countries' experiences. The international growth of the service sector along with a boom in flexible working practices have changed the industrial landscape almost beyond recognition. However, the extent to which policymakers have prioritised (or even noticed) employee health is put into question by a range of evidence. It is to this that we now turn.

The changing industrial landscape

Labour market restructuring and OHS

> Late twentieth-century ('turbo') capitalism is characterised by ever more variety in the contractual arrangements under which work is done.
>
> (Cully *et al.* 1999: 300)

On an international scale, both governments and employers have promoted labour-market flexibility and employment flexibility. The OECD has identified the importance of labour-market flexibility as a contributor to economic growth (OECD 1989, 1997). Consequently, the removal of restrictions on the type of contract employers can offer on previous job boundaries, and the introduction of new working patterns, sub-contracting and flexible reward systems, have proliferated throughout Europe (see, for example, Blossfled and Hakim 1997; Felstead and Jewson 1999). In turn, flexibility has been seen as playing a potentially important part in achieving various organisational objectives, including lower labour costs, improved responsiveness to market uncertainties, greater utilisation of plant and equipment and higher-quality output (Blyton and Turnbull 1995).

The growth of flexible working has been a key feature of the UK labour market in recent years. The temporary workforce in the UK has grown steadily in recent years to a position where there are now around 1.75 million temporary workers, representing more than 7 per cent of all employees. This compares to around 5.5 per cent a decade ago (IRS 1999). According to the Labour Force Survey figures, the number of temporary employees in the UK grew by around 30 per cent between 1995–6, compared with an increase of only about 2 per cent in the total number of employees during the same period. In addition, the number of temporary agency workers increased by a massive 148 per cent (Labour Market Trends 1997). Employers are increasingly adopting such arrangements to meet fluctuating demands for labour in the short term, and to adapt to technological and organisational change in the longer term. However, according to one report, employer abuse of flexible working arrangements can lead to deteriorating working conditions and insecurity of employment (IRS 1997: 2). The report finds that 'too often flexibility is a one-sided bargain, with workers expected to be at the beck and call of their employers, but employers making few attempts to accommodate the needs of their staff'. The 'one-sided bargain' is not unique to the UK. In New Zealand, McLaughlin and Rasmussen (1998) found that in the retail sector many employees are required to work flexible hours and days but are unable to exercise much influence over those hours and days of work. It has also been suggested that employers use flexibility as a means to cut costs and minimise their legal obligations to the workforce via an increasing use of agency workers and outsourcing (IRS 1997: 2). In addition, the same report finds that flexibility increased gender pay differentials, particularly with overtime and shift-work being assigned to men and temporary and part-time work to women. Indeed, UK figures for the first part of 2001 show that the number of people in part-time employment was 7.1 million. Of this total, 1.42 million were men and 5.59 million were women (Labour Market Statistics, June 2001).

It is only recently that international governments and regulatory agencies have begun to legislate for the growing population of temporary, part-time, home and agency workers, despite the growing body of evidence that suggests the negative impact on worker health and safety in the deregulated, restructured labour market. The problems in OHS legislation are not just about coverage of different employment groupings, but also of enforcement activity. Quinlan (1999: 444) argues that while OHS laws purport to cover all workers, little compliance activity has been directed to smaller workplaces, subcontractors or more casualised forms of employment. Consequently, it is likely that tribes of workers orphaned from national OHS regulations and enforcement activities will be found in most industrialised countries.

While some changes to labour markets may have had positive effects on injury and illness rates, such as the shift in employment away from traditionally hazardous industries, there is limited knowledge regarding the patterns of injury and illness in the growing service industries, where the concentration of temporary and other contingent workers (for example, self-employed, homeworkers) is particularly high. However, under-reporting and the fact that many jobs are in the grey economy (and are therefore difficult to examine) are key obstacles in developing a realistic picture of OHS. Quite worryingly, Quinlan (1999) reports how contingent workers and those workers who believe their jobs are insecure are less likely than other workers to report accident/injuries or illness of greater concern, which has ramifications for policymakers who are using essentially inaccurate statistics to direct future policy initiatives and interventions.

Notwithstanding the possibility of missing data, a growing body of international evidence points to a higher incidence of serious injuries in industries where workers are overwhelmingly employed on a temporary and

casual part-time basis (Quinlan 1999: 440). In a review of fifty-nine international studies (covering eleven countries) published between 1982 and
2001, Bohle, Quinlan and Mayhew (2001) report that over 90 per cent of
these studies found job insecurity to be associated with a significant deterioration in health and well-being. This is a matter of great concern given that
on an international level, workers report an increase in job security (OECD
1997).

It is not only downsizing trends that can create job insecurity amongst
employees, but also the changes to employment contracts and broader
employment practices, which weaken job tenure (for example, subcontracting). Johnstone, Mayhew and Quinlan (2001) cite a range of international
evidence that underlines the negative impact of outsourcing on OHS. In a
Finnish study of ninety-nine serious accidents, Salminen *et al.* (1993: 356)
found that subcontracting increased the accident risk one and a half times
(quoted in Quinlan 1999: 433). Comparisons between outsourced and non-
outsourced workers in building, garment making, hospitality and childcare
found that OHS risks were worse for outsourced workers and their presence
alongside permanent employees also exacerbated OHS risks more generally
(Mayhew, Quinlan and Ferris 1997; Mayhew and Quinlan 1997a, 1997b,
1999). The risk factors associated with outsourcing are summarised in Table
2.1.

Other studies report that downsizing is linked to higher absence rates,
especially amongst older workers (Vahtera, Kivimaki and Pentti 1998;
Szubert, Sobala and Zycińska 1997). Absence figures, like compensation-
claim statistics, are not definitive or necessarily complete OHS indicators.

Table 2.1 Risk factors associated with outsourcing

Economic and reward factors
- Competition/underbidding of tenders
- Taskwork/payment by results
- Long hours
- Underqualification and lack of resources
- Off-loading high-risk activities

Disorganisation
- Ambiguity in rules, work practices and procedures
- Inter-group/inter-worker communications
- More complicated lines of management control
- Splintering of OHS management system
- Inability of outsourced workers to organise/protect themselves

Increased likelihood of regulatory failure
- OHS laws focus on employees in large enterprises
- OHS agencies fail to develop support materials
- OHS agencies fail to pursue appropriate compliance strategies

---

Source: Quinlan (1999).

For example, some employees fearing redundancy may avoid absence or making compensation claims for fear that it may affect their future employment in the firm or other work contracts (Quinlan 1999: 438). The extent to which 'insecure' workers and workers in general are under pressure to maintain 'clean' absence records is highlighted by a 1995 TUC report, which found that a high majority of UK employees are afraid to take time off work when they are ill.

An interesting illustration of patchy employment protection and related negative OHS outcomes (as well as offering a tribute to free-market logic) is provided by the Australian long-distance trucking industry. This industry was deregulated in the 1970s and is infamous for its poor, and worsening, safety record. Truck driving is classified as one of the most dangerous occupations, and these risks extend to other road users (Quinlan 2002: 2). In 1999, 189 Australians (fifty-one of whom were truck drivers) died in crashes involving articulated trucks, representing just under 10 per cent of all vehicle fatalities. Compared to citizens in the USA, Finland and the UK, available evidence suggests that Australians are almost twice as likely to die in a crash involving an articulated lorry (Quinlan 2002: 2). Evidence dating back to the mid-1980s identifies a range of serious OHS risks related to commercial practices (for example, excessive working hours, reward schemes) and to ineffective and inadequate regulation and enforcement activity. These elements combine to produce a volatile and unchecked safety environment complete with a minefield of OHS hazards. Despite calls dating back to the 1980s from trade unions and other employee-representative groups, operating licences are not required unlike for other modes of transport such as railways. In embracing all operators, including the reckless and irresponsible, competition intensified, freight rates lowered and fatalities and crashes escalated. Equally, the occupational health of truck drivers has been of serious concern. Quinlan (2002) reports a high incidence of illness and injury amongst a representative sample of 300 Australian long-distance truck drivers. The study found that over half of the drivers reported chronic injury, such as a back injury or hearing loss. In addition there was evidence of drivers experiencing considerable stress (based on General Health Questionnaire results), a finding that may go some way to explain the particularly high suicide rate amongst drivers.

It is not only the inadequate coverage of regulation that contributes to the poor safety record in the industry, but also its focus. According to Quinlan (2002: 13), past intervention has failed because it focused on symptoms of the problem (for example, speeding, driving hours) rather than the commercial practices that encourage unsafe behaviour (for example, scheduling pressures, unpaid waiting time, job insecurity). In other words, the bulk of responsibility lies with the drivers without giving due attention to employer behaviour. For example, the onus is on truck drivers not to drive while fatigued or to break speed limits, yet legally Australian truck drivers are allowed to work up to 72 hours a week. The continuance of bonus/

penalty systems in relation to delivery time and scheduling is likely to be an exacerbating factor. We can see, therefore, the interaction between macro- and micro-level factors, namely legislation and workplace factors such as economic incentives.

Responses by policymakers

For policymakers, subcontracting presents some key OHS concerns. According to Quinlan (1999), a purely economic approach to labour market changes has led to the wider consequences of these changes for workplace health and safety. The allocation of clear lines of responsibility, the effectiveness of monitoring adequate safety and health procedures and the scope of coverage of employees are made increasingly difficult by the elastic nature of contemporary definitions of 'the workplace'. The ubiquity of subcontracting, fixed-term contracts and temporary 'agency' workers makes it equally difficult to identify an organisation's 'workforce'. A key problem is the lack of inclusive regulations that identify the locus of legal responsibility for proper procedures of safety management within complex patterns of contractual agreements. To a significant degree, OHS legislation in a number of countries is still directed to single-entity employers and their employees in large workplaces, thus failing to incorporate new, more complex, patterns of employment. Arguably, the response of regulators and policymakers has been largely reactive, for example, in responses to major incidents where a causal link has been made with subcontractors (for example, the Valujet crash[7]). Subcontractors were also implicated in the UK Hatfield railway crash in 2000, with reports that Railtrack had questioned the standards of a major track repair scheme carried out by subcontractors on the Hatfield line only two weeks before the derailment (*Guardian*, 25 October 2000).

Under British and Australian regulations, employers cannot use outsourcing to formally avoid their liability under general duty provisions, which may be a revelation to some employers (Johnstone, Mayhew and Quinlan 2001). However, the general weaknesses of the provisions means that the complexities posed by subcontracting arrangements are often not fully appreciated or addressed by employers. Positive moves are, however, evident in Australia, where in recent years a number of states have introduced policies requiring greater co-ordination of the activities of subcontractors and contractors with the main employer. However, the extent to which these policies will be adequately enforced and monitored is put into question by the shrinking resources available to the regulatory authorities. We discuss this further in the next chapter. For now, we travel to the final stop in our whirlwind tour around the macro-level influences on OHS policy: the role of societal pressures for change.

Globalisation and social pressures

Changes in expectations and values – some of which will be reflected in national and international codes and standards from, for example, the International Labour Organisation (ILO) and World Health Organisation (WHO) – are influenced by a wide range of factors, including economic development, levels of prosperity and the level of information about policy and practice in countries across the globe. Advanced industrial economies are increasingly utilising Second and Third World countries to expand manufacturing and other operations (for example, data-processing/call centre work), while increasing cost efficiency. Facing powerful economic and social constraints, Second and Third World countries can be tempted to take on industrial processes and patterns of employment that are no longer tolerable in the advanced economies. For example, high-hazard industries such as coal mining are now predominately located in eastern Europe, Africa, China and India. In 1980, for example, UK private-sector manufacturing and extraction industries accounted for 25 per cent of all workplaces and 38 per cent of UK employees. This figure had fallen to 18 per cent and 32 per cent respectively by 1998 (Cully *et al.* 1999). By contrast, there was a 4.9 per cent growth in mining and 3.6 per cent growth in manufacturing in India over the period 2000–2001 (IndiaOneStop.com 2001). With this in mind, the reductions in workplace fatalities and the slow-down in major injuries in many of the developed countries is hardly surprising. While workers in the advanced economies may benefit from carrying out less hazardous work as well as enjoying greater levels of employment protection, the opposite may be true of Second and Third World countries.

This export of hazard ranges over health risks associated with asbestos, pesticides and hazardous waste (especially nuclear waste). It appears that capital can, with relative ease, export hazard to countries with lower regulatory standards, weaker labour resistance and compliant governments anxious to attract inward foreign investment (for example, Union Carbide in the Bhopal chemical plant disaster). One poignant example is that of the Chinese government, who, throughout the trial of a Shenzen toy-manufacturing company where eighty-four young girls were killed and a further forty-five disabled following a fire in the factory, was reported to have continually stressed the importance of not undermining the confidence of foreign investors (quoted in Nichols 1997: 108).

Nichols (1997: 106) argues that multinational companies (MNCs) bear a major responsibility for the lower standards of employee health protection evident in the developing countries as part of capitalism's world division of labour. In this context, workplace safety standards are often shaped by specific and variable national regulations, for example, the requirement for guards on machinery and the issue of safety equipment such as hard hats and hearing protectors. Such differences were noted by Wokutch (1990) in a study of the Kenyan division of an MNC. The author argues that only a disaster or a significant deterioration of OHS would seem likely to bring about increased attention and commitment of resources on the part of the parent company.

However, increasingly, labour and environmental standards of major multinationals are under intense scrutiny from global consumer and environmental groups. A brief glance at the websites of campaigns devoted to such brand names as Nike, GAP, Nestlé, Monsanto, or of those dedicated to challenging the large oil companies, suggests that there are business risks involved in failing to meet high standards of corporate ethical behaviour. There is growing pressure on companies from international organisations such as the International Labour Organisation, and from international trade-union confederations, to adopt appropriate 'business' or 'social' ethics, culminating in the appearance of numerous social codes and voluntary agreements in industry. The issue of ethical trading has gained prominence in recent years, as consumers become more aware of employment practices across the world. Consequently, they are now beginning to demand that products they purchase have not been manufactured under conditions that violate human rights, for example, in the use of child labour.

In France, the FCD retail commerce employers' federation campaign, 'Social Clause', was launched in 1998, which provides social codes of practice. The FCD's campaign is mirrored at the European level, where on 6 August 1999 an agreement on fundamental rights and principles at work was concluded between European employers' organisations. Under the terms of the agreement, organisations must abide by and promote key rights affecting health and safety, actions to abolish child labour and respect the right to the freedom of association and collective bargaining (*European Industrial Relations Review* 324: 13–22, 2001). In addition, the framework European Directive, 'Equal Treatment in Employment and Occupation', was formally adopted on 27 November 2000 and is expected to be transposed into national member states' legislation on or before 2 December 2003.[8] In this example, consumer power drove policymakers to formulate regulations, but the consumers' power was derived from their ability to hit capital where it hurts – boycotting their products and/or circulating bad publicity. The growth of consumer power is an interesting topic and its ability to affect capital and government will be heavily dependent on the ability to mobilise and engage collective groups of consumers. How will governments and employers respond to challenges to their prerogative by the hand that feeds? If recent headlines are anything to go by, activists will be labelled anarchists and terrorists, bearing some similarity to the depiction of trade union activists during major industrial action in the 1980s (for example, the UK mining and newspaper industries). In all of this, the combined force of community action, legislation and employee involvement is clearly needed to effectively oppose the will of capital in what appears as an all too common under-prioritisation and abuse of employee health and safety.

Conclusions

This chapter has discussed a number of macro-level influences in the regulatory politics of OHS, including political regimes, deregulation, privatisation and labour market restructuring. From this we have observed how the dominant ideology of free market economics – bathed in cost rationalisations and instituted by cost–benefit analyses and risk-assessment exercises – appears to operate against positive OHS outcomes. As the UK's HSE come to terms with cost–benefit analyses (out of necessity rather than choice), the extent of their assimilation to government ideology is evident in their proposition that increased risk (for workers) is deemed acceptable as long as 'substantial benefits (including economic savings)' are made (HSE 1999a: 74). This presents a gloomy picture for employee health and safety, particularly when we consider the hazy concept of 'tolerable risk', where a key question is 'tolerable to whom?' As we will see in the next chapter, it appears that increasingly management is dominating the identification or risk. James and Walters (2002: 149), for example, report on a 1998 TUC survey of safety representatives, which suggests that safety representatives have only a limited involvement in risk assessment. Fewer than one in three respondents were satisfied with the extent of their involvement in risk assessments, while two in five said they had not been involved at all. The main point here is that the difficulties relating to how risks are perceived, identified and measured, as well as the extent of workers' freedom of choice to accept risks in the workplace, all combine to produce poor odds for workers in the workplace health and safety gamble.

Overall, an economic approach to OHS reduces the social dimension of the workplace environment to simple quantitative values, leaving out the connections between safety, social relations at work and the larger balance of power between workers and employers in society. According to Dorman (1996: 188) the technical 'fixes' endorsed by the risk assessment industry target a certain level of safety but do not alter the social or strategic environment within which risks arise leading to limited (if any) real improvements in workplace safety. Positioned as a cost variable in risk assessment and cost–benefit analysis exercises, OHS is a hostage to fortune. As critics have argued,

> Theories favouring the private market for safety are based on unfulfilled assumptions . . . firms frequently do not have sufficient knowledge about hidden costs of accidents to enable proper cost–benefit accounting. Nor do they typically have enough experience with fatalities to be able to take preventative measures.
>
> (OECD 1989, quoted in Moore 1991: 13)

Emerging from the discussion is the crude issue of power – who dictates, who decides on what is acceptable and allowable, and is there anyone able to challenge these decisions effectively (and without facing victimisation)? One might argue that since employers and employees have 'equal' interests in the area of health and safety, both parties should have equal influence in decisions affecting health and safety at work. This further highlights the importance of accounting for power relations and inequalities in the employment relationship. Even where there is a general consensus that work should take place in a healthy and safe environment, the countervailing pressures for the production of goods or the delivery of services continually test management's commitment to their legal obligations. The potential danger is that productivity and profit imperatives will be allowed to take priority over all other issues. That this is a common reality simply mirrors the fact that too often capital is able to award itself a licence to define, manipulate and create risks. Moreover, it could be argued that reprimanding capital for doing so aligns to scolding a child for playing with a toy he or she has just been given.

With all of these issues in mind, what potential does HRM hold for delivering the conditions most appropriate for 'good' workplace health and safety management? As outlined earlier, at least two prerequisites are identified, namely, a high level of employee involvement in health and safety and a degree of consensus over health and safety issues. Both of these require suitable mechanisms and support. The question is, to what extent is HRM capable of providing these? We consider this in detail in the next chapter, where we explore the extent to which workplace safety is (or if it has ever been) an area of 'general consensus' between employers and the employed. This is a crucial point given the emphasis placed on effective partnerships by the UK government. For example, the 'revitalising' strategy on OHS states: 'Effective partnerships between all stakeholders in the health and safety system . . . are crucial' (DETR/HSC 2000: 18).

The building blocks of effective partnerships may include employee involvement mechanisms such as joint consultative committees. Like any partnership, be it friends, family or work colleagues, differences in perspectives are likely to emerge at some point. The means to resolve these conflicts form another crucial part of securing and sustaining 'effective partnership'. However, is HRM up to the job?

3 The social processes of OHS

By some accounts, 'a greater natural identity of interest' exists between employers and the employed over OHS, and because of this, conflicts of interest are less likely to arise. Such assumptions about consensus in the employment relationship fit well with the ideology of HRM and they are, to some degree, supported by the statistics on industrial (strike) action in the UK. In 1998, the number of stoppages (strikes) in the UK plummeted to the lowest level since record-keeping began in 1891. Other data shows that across twenty-three OECD countries, the UK had the eighth-lowest strike rate[1] (TUC 2000). Compared to other European countries, the UK had 12 days per 1,000 employees, in contrast to Spain with 127, Italy with 40 and Ireland with 31. The US figure was 42, while in Australia the figure was 78. Based on these figures, industrial harmony in the UK might appear at an all-time high. However, strike action is not the only expression of industrial unrest. A range of other statistics put into question the dubious proposition that employers and trade unions have miraculously reached a state of consensus, particularly those relating to the exponential rise in the number of ballots for industrial action in the UK.

A Trades Union Congress (TUC) study of ballots for industrial action (covering 81 per cent of the TUC's affiliated membership) reports that 60 per cent of affiliated unions had organised ballots between June 1999 and May 2000, and the majority (57 per cent) had done so on five or more occasions. For that period, the unions had organised 983 ballots for industrial action – nearly twice the number organised in the previous year (TUC 2000). With only 32 per cent of ballots leading to industrial action, unions appear to be using these as a (very expensive) bargaining tactic. However, expensive with the average cost of ballots reported to be as high as £18,747 (TUC 2000) this approach to resolving industrial disputes may not be open to some of the smaller trade unions. In the most recent survey of this type, this cost had risen to £32,700 (TUC 2001a) – a hefty price shouldered by the trade unions and their members to achieve agreement over so-called consensual issues such as health and safety.

Waddingon and Whitson (1996) report how health and safety remains one of the most common issues raised by union members. Indeed, just over one-third of active trade union members surveyed referred to health and safety as one of their three most common grievances. This trend was also evident in the WERS (1998), where health and safety was one of the top issues dealt with by trade-union and employee representatives (Cully *et al.* 1999). In terms of industrial ballots, along with pay, changes to working practices and redundancies, health and safety was one of the most common reasons for ballots in 1999, 2000 and 2001 (TUC 1999a, 2000, 2001a). All of this goes some way to invalidate any assumptions about consensus in the employment relationship, and also raises questions about the performance and impact of a range of HRM policies and practices that, as we saw earlier, are highly popular in the UK.

This counter perspective on consensus is further supported by the incidence of employee tribunals in the UK. Over a ten-year period (1990/1 to 2000/1), the number of employee tribunals has risen threefold, from 43,000 in 1990/1 to 130,000 in 2000/1. However, rather than addressing the causes of these expressions of discontent, the Labour government has instead chosen to introduce measures[2] that are likely to inhibit employees from taking this route of redress in the future, arguing that the rise in the number of tribunals is an unacceptable cost burden to the taxpayer (*Observer*, 5 August 2001: 16). The government's position on employee rights is made a little clearer by their intervention, which is only likely to fuel the growth of power inequalities within the employment relationship. This brings the power relations between employers and the employed into focus.

In this chapter, we focus on some of the actors involved in the 'workplace health and safety partnership'. We consider the role of trade union organisations, enforcement agencies and employers and the processes and channels used to facilitate 'good practice' in OHS. Representative structures are of primary concern since these provide the forum for employees to voice their interests in the process of negotiating and securing consensus. While regulations are in place in the UK that require employers to consult with employees over health and safety issues, the extent to which these structures lead to meaningful employee involvement in OHS decision-making processes is, by some accounts, highly questionable. In reviewing these areas, we can gauge the extent to which HRM principles and practices, within current political and economic climates, provide optimal conditions for effective health and safety management.

Trade unions and employee involvement

The Labour movement, founded on trade union organisation, has played a major role in addressing health and safety at work throughout the twentieth century. It was through the trade union form of collective organisation, particularly in the nineteenth century, that many workers began to seek relief from danger and work-related illness. It was out of these concerns that many in the labour movement, unions and their parliamentary counterparts campaigned for legislation to address ill-health and danger at the workplace, as well as for the welfare provisions that have been elaborated in many societies during the twentieth century (Fairbrother 1996). The harsh reality is that OHS is not, and has never been, characterised by shared interests (Carson 1985, 1989; Carson and Henenberg 1988).

The intensely political nature of OHS is underlined by the correlation between the relative strength and ability of capital and labour in the employment relationship, and the UK industry's safety performance in the 1970s and early 1980s. While flashbacks from these eras may include images of violent clashes between picketing trade unionists and the police force, widespread strike action and Margaret Thatcher's determination to disable the trade unions that 'were crippling the British economy', one truth might be overlooked. According to Nichols (1997), during the high point in levels of influence during the 1970s, a range of employment protection regulations and provisions were put in place. Indeed, employment legislation (including regulations covering OHS) was most robust during periods of strong trade union influence in the 1970s, while the subsequent dilution (and deletion) of employment protection and employee rights followed a linear pattern of decline with trade union power during the 1980s and 1990s. The destructive impact of anti-union legislation, biting recession, swells in unemployment and sectoral shift, cumulatively eroded trade unions' power, role and scope during this period. These developments have significant implications for safety representation.

Under the SRSC Regulations 1977, unions were awarded significant rights that enabled a meaningful level of participation in workplace health and safety management. James and Walters (2002: 147) emphatically point out that these regulations were not an outcome of the Robens Committee (1972) recommendations, and were instead the result of a long trade union campaign that dates back to the 1960s. Indeed, the rights contained in the SRSC regulations were far in excess of the minimalist consultation provisions put forward by Robens. The SRSC regulations provide extensive rights, including, for example, the right to carry out workplace inspections, to investigate dangerous occurrences/ accidents, and the right to be involved in external inspections by the regulatory bodies. However, since then, a substantial decline in trade union density has occurred leaving in its wake what has been termed a 'representation gap'.

Since the SRSC regulations only relate to unionised workplaces, patchy coverage began to emerge as union density levels declined throughout the 1980s and large unionised workplaces were increasingly being replaced by smaller, non-union enterprises. By 1998, 47 per cent of workplaces had no union members at all (an increase from 36 per cent in 1990), while only 2 per cent of workplaces were fully unionised (down from 7 per cent in 1990) (Millward *et al.* 1992; Cully *et al.* 1999). In addition, small enterprises were characterised by an absence of both recognised unions and consultative committees (Cully *et al.* 1999).[3] The Health and Safety (Consultation with

Employees) (HSCE) Regulations (1996) set out to provide cover for the representation gap that had emerged, requiring employers to consult employees in non-union workplaces, crucially, either through a representative or directly. However, the new regulations did not provide a panacea for the representation gap, but instead were instrumental in the further curtailment of workers' representation rights.

This charge is based on two key features of the regulations: the limited range of rights for safety representatives in the HSCE regulations and the focus on direct communication methods. First, the HSCE regulations do not award safety representatives the same rights as those included in the SRSC regulations, for example, the right to workplace inspection or the right to set up a joint safety committee (for a full discussion of these regulations, see James and Walters 2002). Second, if employers choose the alternative route of direct communication, what may result is a simple one-way communication exercise as opposed to allowing any meaningful worker involvement in health and safety issues. In the WERS (1998), over half of employers reported that they chose to consult directly with employees using mechanisms such as e-mail, briefings, newsletters and noticeboards.

While the apparent haemorrhage of rights for safety representatives along with the dubious value of direct communication with workers have generated concerns that health and safety has been captured by management, the continued popularity of joint safety committees (employer and employee representatives) offers some prospect of two-way communication and consultation. Cully *et al.* (1999) report that 39 per cent of all UK workplaces surveyed operated joint health and safety committees and not suprisingly, given the differences in representation rights between the SRSC and HSCE regulations, joint safety committees were most common in workplaces with union recognition. One caveat is, however, that historically joint safety committees are found to be more effective when they operate through established trade union channels (Beaumont *et al.* 1982). This in turn generates some concern over their effectiveness in non-unionised firms. In all of this, a key concern is that the prevalence of direct consultation mechanisms and the propensity of management to appoint safety representatives may mean that, in some cases, contemporary consultation forums may be no more than puppet shows run by management.

Perhaps in recognition of some of the weaknesses in contemporary approaches to employee involvement in OHS, the Health and Safety Executive (HSE) make a clear statement on their preferences. In their words:

> We (the HSE) would far rather see safety representatives. There's no doubt about it. They add value and so if there is something that actually promotes the appointment or election of safety representative rather than relying on direct consultation with workers, that we think would be value added.
>
> (Jenny Bacon, Director General, HSE, January 1998)

The value that safety representatives are actually capable of adding is, however, undermined by the HSE's reluctance to enforce the rights given to employee representatives. As James and Walters (2002: 149) explain, the HSE has long avoided inspector involvement in cases where employers are reluctant or unwilling to consult with safety representatives or fulfil any of the other requirements in the relevant legislation.

With fewer unionised workplaces, the stifling effect of consultation regulations on meaningful employee involvement in workplace health and safety, and the popularity of direct communication, it is unsurprising that by 1990 management alone determined health and safety in about four out of ten establishments (Nichols 1997). Why should this present a problem? Nichols (1997), amongst others, might argue that contrary to the simplistic assumption that health and safety is a shared concern and an area of consensus between employers and employees, OHS is determined on a political and economic basis. Behind the rhetoric of 'safety makes sense' is the rudimentary issue of power. As such, management's apparent domination of OHS agendas, particularly in non-unionised firms, is indicative of the wider trends of growing power inequalities between employers and the employed. It would, however, be unreasonable to assume that meaningful consultation can only take place when there is trade union representation, although as already mentioned, there is substantial evidence to show that higher injury and illness rates occur in workplaces where management alone decided on health and safety issues (Walters 1996; Millward *et al.* 1992; Reilly, Paci and Holl 1995; James and Walters 1997; Litwin 2000).

In all of this, a key point of importance is that recent developments in the UK run counter to historical patterns and lessons in relation to improved health and safety outcomes, meaning that the goals set by the HSE and the government in relation to OHS improvements may be far out of reach under present conditions. Rather than equipping workers with the means to improve workplace safety, it appears, that their rights have been continuously eroded. Unfortunately, the UK government's attempts to airbrush over the ugliness of power inequalities and opposing interests, as well as the limitations and weaknesses of a self-regulation, may mean that the OHS agenda will remain subject to the whims and convenience of capital. However, for James and Walters (2002: 154), legislative reforms could do much to reverse this situation. The authors suggest a variety of measures that would lead to the enhancement of the rights and functions of safety representatives or, alternatively, they suggest the creation of a general framework of worker representation that encompasses the issue of health and safety.

While trade unions may occupy a fundamental role in the regulation of workplace health and safety, the responsibility for monitoring and enforcing OHS legislation lies with the enforcement agencies. Given that it was estimated that UK workplaces will be visited on average only once every seventeen years, compared with every four years two decades ago (Labour

Research, April 1997), our expectations of the monitoring capabilities of the enforcement agencies might be rather low. In response to a similar state of affairs in Australia, some trade unions have taken radical steps. Because of a perceived lack of government inspection activity, one Australian trade union initiated its own prosecutions for breaches of the regulations covering homeworkers' working terms and conditions (Johnstone, Mayhew and Quinlan 2001). The extent to which similar action emerges in the UK will be of interest.

Government enforcement agencies

In the UK, following the introduction of the HASAWA (1974), the roles of the HSE and the tripartite HSC were established as being independent from other policy exigencies, while significant government support was provided through funding and other resources. However, as alluded to in the last chapter, the role and scope of the HSC/E has been constrained considerably since then following the imposition of a 'business-friendly', 'partnership' approach by successive UK governments and drastic cuts in resources during most of the 1980s and 1990s (Bain 1997, Woolfson and Beck 1998). In the Health and Safety in Local Authorities (HELA) Annual Report (2000) the 'particular concern over the reduction in resources allocated by local authorities to health and safety enforcement is noted'.[4] The concerns are well founded given the impact on the number of inspectors employed – a 16 per cent fall in the number of full-time inspectors from the previous year to 1,210, making it the lowest level since 1991/2 – and the number of visits by inspectors reported as being down 23,000 from 1997/8. These cuts have been made despite a steep increase in the number of 'over-3-day injuries' and fatalities in local authority enforced sectors (HELA 2000: 3). These trends are repeated in the following year. In its 2001 Annual Report, HELA reports a continued decline in the resources devoted to local authorities' health and safety enforcement activity (25 per cent in the past five years), meaning fewer health and safety inspectors were employed (down by 8 per cent, making the number of inspectors the lowest since 1986/7), and subsequently fewer workplace inspections were carried out (19,000 fewer than the previous year) (HELA 2001).

These trends appear to mirror those in the USA. During the early 1980s under the Reagan administration, the Occupational Safety and Health Administration (OSHA) – the US equivalent of the HSE – along with other forms of 'social' regulation came under political attack. After slashing OSHA funding and the size of the inspectorate, the Reagan administration devised new procedures to improve the efficiency of its remaining force. Dorman (1996: 193) reports how in 1981 it became OSHA policy not to respond to worker complaints over safety, while firms with above-average safety records were exempted from on-site inspections (Abel 1985), creating a clear incentive for firms to under-report injuries and illnesses. As a result many large firms, including Union Carbide, Chrysler and General Motors, have since been issued fines for wilfully under-reporting accident data (Risen 1987; Karr 1987). We can speculate about the position of OHS in the federal government's agenda when we consider Dorman's (1996: 193) observation that the federal government employs more game wardens to protect wildlife from illegal hunting than safety inspectors to safeguard the lives and health of US workers.

Highlighting employers' reluctance to implement legislation voluntarily, and, by implication, the dubious efficacy of self-regulation, is the rise in the number of formal enforcement notices issued by UK inspectors. Despite fewer visits by inspectors, formal enforcement notices rose by 44 per cent from the previous year, representing a rate of 5 per 1,000 premises. This trend is mirrored in statistics covering all sectors (HSE 2000c). With the exception of the construction industry, the numbers of notices and convictions in every other key industry have continually risen from 1997 to 2000, as shown in Table 3.1. It could be argued that these figures represent only the 'tip of the iceberg', given the potentially masking effect of a 'business-friendly approach' by the HSE, never mind the limits that shrinking financial and staff resources place on enforcement activities.

Supporting minimal interference from enforcement agencies is the affordable level of safety fines. The overall success of the HSE and the courts in adopting the government-imposed 'business-friendly' approach to OHS is evidenced in the low numbers of pejorative penalties imposed on law-breaking organisations, as shown in Table 3.2. This shows that in a 12-month period, more than half of the penalties are for £2,500 or less (57 per cent), and that over three-quarters (77 per cent) are for £5,000 or less despite 40 per cent involving fatalities. In addition, an HSE report states that the average fine during the period 1 April 1999 to 31 March 2000 was £4,597 (this excludes twelve fines of £100,000, which would raise the average to £6,744) (HSE 2000c).

While apparently content to continue with a minimalistic, self-regulatory approach to OHS in most industries, a recent HSE discussion document explores the possibility for imposing more detailed and rigorous requirements on duty holders in 'high-risk' industries through permissioning regimes (HSE 2000d). Permissioning regimes have been introduced in various 'high-risk' industries following specific accidents, such as the 1957 Windscale (nuclear power plant) fire and the 1988 Piper Alpha disaster. According to the HSE, the existence of these regimes reflects society's concerns about risk in specific industries and the need for an added level of confidence. This is secured through a structured and documented approach that goes beyond the more general duties laid out in the HASAWA (1974). 'Extras' include the advance provision of systematic assessment in lifecycle hazards and risks, from design to decommissioning, where any reduced risk values shown by quantitative risk assessment (QRA) must be supported by a clear and satisfactory description of the practical measures that deliver

Table 3.1 Enforcement activity by the HSE 1997–2000

Year	Agriculture, forestry, fishing	Extractive and utility supply industries	Manufacturing industries	Construction	Service industries
Notices laid					
1997/8	80	32	518	719	278
1998/9	117	49	601	631	311
1999/2000	262	90	716	796	389
Convictions					
1997/8	69	26	438	544	207
1998/9	102	34	551	565	260
1999/2000	103	62	603	537	297

Source: HSE (2000d).

Table 3.2 Fines by HSE for the period 6 April 1999–11 May 2001

Fine	Number (n = 565)
£0	43 (12 cases had related breaches which incurred fines between £2,000 and £400,000)
£1	1 (this case had a related breach which received a £74,999 fine)
£2–£100	10
£101–£250	34
£251–£500	44
£501–£1,000	64
£1,001–£1,500	32
£1,501–£2,500	109
£5,001–£10,000	57
£10,001–£20,000	36
£15,001–£25,000	24
£25,001–£50,000	2
£50,001–£100,000	4
£100,001–£200,000	4
£200,001–£399,000	0
£400,000	1
Over £400,000	1 (£1,500,000 GW Trains in relation to the Southall crash)

Source: Figures taken from the HSE website (www.hse.gov.uk) on 10 July 2001.

them. In addition, regular reviews of risks, hazards, risk control and miti-gation measures are required and, crucially, the entire process must be audited. Greater involvement of the safety regulator and increased trans-parency and accountability of the 'duty holder' are defining features of this approach.

However, another defining feature is the goal-setting approach. Most of the health and safety law currently applied in permissioning regimes and elsewhere is goal-setting rather than prescriptive. This means that while requirements are stated, the methods to achieve them are not. The HSE argues that goal-setting requirements are more likely to lead to tailor-made solutions that suit specific circumstances, assuming, of course, organisations possess the required skills and knowledge to produce such solutions (or can afford to buy it in from external consultants). However, as the HSE notes, industries do not always adopt such an approach, underlining the need for third-party support and intervention – all of which would require significant resources. While permissioning does not guarantee safety, it could go some way to ensuring responsible and effective solutions are developed and imple-mented by duty holders under strict monitoring conditions such as in the nuclear power industry. However, it is important to note the reactive nature of this intervention, where stringent procedures have only been introduced after industrial tragedies have occurred.

One might wonder why this enhanced regulatory intervention is only applicable to 'high-risk' industries. The HSE allude to the issue of resources.

In their words: '[Permissioning] regimes are resource intensive for both the regulator and the regulated ... for this reason "permissioning" is used sparingly' (HSE 2000d: 8). It is cold comfort that employers are reined in only after industrial tragedies such as Piper Alpha, Southall, Ladbroke Grove and Windscale.

For the industries not (yet) classified as being 'high-risk', one published strategy statement, *Revitalising Health and Safety*, sets the course for the management of OHS in the UK (DETR/HSC 2000). This joint government/HSC initiative endorses the basic framework provided by HASAWA and sets out the economic business case for good practice. It also states targets for reductions in work-related injury and illness as well as a range of suggestions for ways to help realise these targets. However, any excitement about the five- and ten-year targets is soon quenched by the fact that they fall well below current trends, providing easy targets for the government to meet. For example, figures for both fatalities and major injuries for the period 1997/8 and 1998/9 show a 9 per cent decrease (HSC 2000a). If these annual decreases were sustained over five years, the government's set target would be substantially exceeded regardless of whether the figure was for 2004/5 or as a cumulative figure for the five-year period. As such, the government appears happy to stand still – a position well hidden in the pompous rhetoric.

The government's intention to intervene as little as possible in OHS is clearly articulated in the 'Revitalising Health and Safety' document, placing the onus instead on employers, whose responses and actions (or inaction) are likely to be heavily influenced, motivated and shaped by economic and competitive pressures. Based on this it could be argued that the 'revitalising exercise' is based on tactics of persuasion, cajoling and buck-passing to employers and insurance companies. And while some innovative ideas are fielded (but not adopted), such as greater powers for safety representatives to impose provisional improvement notices and accreditation schemes for health and safety, the document delivers an action plan that is fundamentally based on a 'self-interest model'. As discussed in the previous chapter, this is wholly reliant on employers being able and/or willing to recognise the economic business case for improvements in health and safety, and on insurance companies exerting greater pressure on 'poor performers' (that is, employers responsible for work-related death and injury). Moreover, this approach assumes that employers are able to direct the necessary resources towards OHS improvements. As the DETR/HSC (2000) study reports, cost is one of the key factors that prevents employers from taking further action on health and safety. The study found that 85 per cent of employers reported cost, particularly the cost of health and safety training, to be a major inhibitor to improvements to OHS. One can only marvel at the government's logic in removing public funding of this type of training (that is, the trade union and education grant), which was gradually phased out by 1996.[5]

Nevertheless, the cajoling route to OHS improvements appears to dominate government dialogue. One contingency appears to be based on employers exchanging ingrained perceptions of OHS as a dull, administrative compliance-focused exercise for a new, sexier TQM-esque process based on the notion of continuous improvement and investment, not forgetting cost efficiency. While it is tempting to bludgeon this approach, let us consider the potential of what we will refer to as an 'ownership approach' to health and safety management (HSM). In doing so, we consider the influence of corporate culture, and management style and attitudes, on OHS policy and practice at corporate level. This emphasis on culture, management style and attitudes brings us back to HRM principles, policies and practices, and their potential impact on OHS management.

Corporate culture and management attitudes

Arguably, health and safety is one of the least attractive areas for managers. Cast as the ugly sister of HRM, OHS lacks the glamour and popular appeal of culture management programmes or training interventions. Constrained within a mindset of compliance to legislation and the threat of sanctions, some managers may even perceive OHS as a barrier or a nuisance obstruction to operations. Indeed, studies to date have identified poor communication from line managers over health and safety issues as a major problem in improving OHS outcomes (for example, Vassie 1998; HSE 1987, 1991). Vassie's (1998) study in a multinational chemical company also identified productivity pressures and staff shortages as obstacles to better practice. Adapting managers', and in particular line managers', attitudes towards OHS therefore represents a major obstacle to securing commitment to an 'ownership approach' to HSM.

The compliance/commitment dichotomy is central to HRM (see, for example, Walton 1985; Ramsay 1995), where the benefits of employee commitment and the ways in which to achieve it are well rehearsed (although evidence to show that they actually work is thinner on the ground). If this dichotomy were applied to health and safety, compliance would be likely to produce minimum standards and require high levels of supervision. Commitment, on the other hand, would produce innovation, creativity and superior outcomes. The government's motives for the various 'culture reform' documents become a little clearer – high levels of supervision require large resources. However, just as employers have found it extremely difficult to secure employee commitment to given business objectives the government is experiencing similar obstacles in securing employers' commitment to health and safety.

One illustration of this is UK employers' response to the Working Time Regulations (WTRS) 1998. While many employers complained vociferously about the increased bureaucracy related to the new regulations, little recognition was given to the potential benefits of eradicating a long-hours culture. For many UK employers, the WTRs are seen as creating barriers, rather than offering any potential cost-saving or profit-enhancement opportunities. Equally, many UK employers perceive the required risk assessments under the Management of Health and Safety Regulations (1992) as yet another bureaucratic nightmare, as opposed to an opportunity to identify the risks that cost them money in the short and long term.

The myopic perspective of UK employers on OHS operates in stark contrast to popular views on quality, and in particular total quality management (TQM), defined as:

> the way the organisation is managed to achieve excellence based upon fundamental principles, which will include: customer focus, involvement and empowerment of people and teams, business process management and prevention based systems, continuous improvement.
>
> (European Foundation for Quality Management 1996, quoted in
> Osborne and Zairi 1997)

Several 'experiments' on applying TQM principles to health and safety management report on a range of successful outcomes as well as a range of obstacles. In 1991, the HSE published a guidance document called Successful Health and Safety Management, where the principles of total quality management are applied to health and safety. The document encourages a much wider and more holistic perspective on HSM, where the goals involve satisfying 'the expectations of shareholders, employees, customers and society at large'. A subsequent research report, produced by the University of Bradford for the HSE in 1997, studied the extent to which organisations adopted the core principles of TQM in the management of safety. Based on data collected from twenty-four UK companies, covering a range of industry types and sizes, the overall conclusion was that TQM principles were applied less to HSM than in the core business, except in those businesses where safety was clearly recognised and accepted as a critical factor for the overall success of the business. Consequently, in those companies that applied TQM principles to HSM, there was a high level of identity with, and understanding of, health and safety at executive level, as well as some recognition that a consistent management approach, as required by external quality-assessment bodies, should extend to HSM.

The study also identified the factors that undermined a strategic and business-oriented approach to HSM. These were organised into four main groups: people factors, process factors, organisational factors and external factors (Osborne and Zairi 1997). Leadership style, executive values and the calibre of safety personnel featured as people factors, while organisational factors included an overbearing culture of compliance and a functional approach to HSM that encouraged a singular and limited role for the health and safety specialist. The external factors inhibiting better HSM appeared to be closely related to corporate culture and management attitudes. These included perceptions about the importance of health and safety to business success (see also CBI 1990).

According to these findings, successful HSM, under current regulatory regimes, requires strong support from all levels of management and a greater level of accountability, served perhaps through an external accreditation process. As already mentioned, the introduction of external accreditation systems were proposed by respondents to the DETR/HSE (2000) survey but, rather oddly given the government's aspirations, these ideas were discarded. The DETR/HSE (2000) consultation exercise[6] also reported preferences for greater involvement of the safety regulator across industries, increased resources in health and safety, and greater transparency and accountability for employers, possibly through auditing, accreditation and/or supply-chain initiatives being applied to HSM. In addition, respondents called for greater inspection and enforcement activity, supplemented by harsher penalties. In fact, only 7 per cent of respondents thought current penalties were adequate, which is hardly surprising considering over half of all penalties over one 2-year period were £2,500 or less. However, these preferences are not evident in the government's espoused approach to OHS, leading us to wonder whose interests are actually being served.

By some accounts, meaningful employee involvement is the crux of successful 'continuous improvement approaches to OHS'. This was evident in the aforementioned study by Vassie (1998), During the 15 month, four-phase project, the greatest contribution appeared to come from employees. Throughout the plant, employees took part in attitude and workplace safety surveys, worked collegially with managers within project teams, and undertook training to equip them with the skills to identify safety hazards and how to take corrective action. While successful in achieving most of its goals, it is important to recognise that the programme was wholly dependent on the favour and whim of senior management – the resources, be it time, people or money, could be withdrawn at any time. The success of the project does, however, highlight how meaningful employee involvement in health and safety not only works, but is absolutely essential.

Conclusions

The theme of 'effective partnership' appears not only central to the UK government's approach to OHS, but to the regulation of the employment relationship as a whole. Based on this, the conditions provided by political and economic climates, along with a range of popular management practices (many of which are related to HRM), would be expected to be conducive to achieving this goal. Optimal conditions, we have argued, include a high level of meaningful employee involvement and participation in the management of workplace safety. In addition, the means to resolve inevitable differences in opinion are essential. However, our review to date suggests that current political and economic climates, along with a range of HRM policies and practices, run counter to achieving these conditions.

Strong political and economic undercurrents in workplace health and safety were illustrated by managerial attempts to dominate the OHS agenda via the forums that are involved in the regulation of workplace health and safety, and also in terms of the legislatively induced distancing of trade unions in workplace health and safety management. And, despite the abysmal statistics, the UK government continues to shy away from increased intervention, favouring instead a 'partnership' approach with business, and at the same time distancing the enforcement agencies from their original portfolios while forcing them to kowtow to the prerogative of employers and capital. Amidst all of this, ongoing attempts by the government to change employers' attitudes towards OHS, as evidenced by a range of 'culture-change' initiatives – for example, 'The Costs of Accidents', 'The Healthy Workplace Initiative', 'Revitalising Health and Safety' – mirror the experiences of so many organisations who have embarked upon ill-fated culture change programmes.

While all of the factors discussed so far contribute to OHS policy and outcomes, these may be further affected by an array of workplace factors. For example, in the case of the Australian truck drivers discussed in the previous chapter, commercial practices were thought to encourage unsafe behaviour, such as the influence of scheduling pressures on drivers' propensity to speed or exceed working time limits. This illustrates how work organisation and organisational culture can exert a strong influence on OHS behaviours and attitudes. This is the focus of the next chapter, in which we explore dimensions of work organisation such as working time and performance targets on OHS policy and practice, as well as considering the impact of the working environment on OHS outcomes.

4 Workplace factors in OHS

> The health and well-being of employees is a key factor in the success of any
> business or organisation. Recognising the value of a healthy workplace will
> ensure that staff are 'healthier, happier and here'. Placing these issues at the
> centre of an organisation's concern will help ensure its continuing effectiveness.
>
> (*The Healthy Workplace Initiative*, UK Department of Health, April 1999)

While many organisations' formal policy statements on OHS echo the one
above, it is often the case that workplace factors undermine good practice. A
range of conflicts and contradictions may be created by economic and
competitive pressures, which at a workplace level may translate into a range
of cost-efficiency strategies that are directed at labour. The impact on
employee health may be considerable given that a range of evidence links
factors such as long working hours, high workloads, performance targets and
electronic surveillance to work-related injuries and illnesses. As stated
earlier, the HRM literature gives minimal attention to the interaction
between people management policies on work organisation and employee
health. It is for this reason that we now go on to explore how various
working polices and practices, impact upon employee health and safety.
While economic and competitive pressures will influence work organisation,
the same is also true for the quality of the working environment. Some
commentators have argued that the built environment should not be
regarded as a neutral factor in explanations of occupational illness and/or
disease. According to Bain and Baldry (1995), the built working environment is a critical element of the labour process because it represents
decisions about cost efficiency, the regulation of energy and the level of individual control over the environment, which in turn have significant implications for the health and well-being of employees. For example, 'sick
building syndrome' is intimately linked with a range of cost-efficiency measures affecting building design and construction and policies on work organisation. New technologies occupy a central role in these debates, where their
impact on the working environment and working practices is highly evident
in a range of settings (for example, manufacturing, airlines and call centres).

The preponderance of workplace factors in the OHS equation is underlined by the HSE, who report that in 1995 an estimated 2 million individuals in Great Britain were suffering from an illness that they believed to be caused by their work (current or past). Work-related illness, as considered separately from work-related injury, incurs massive personal and organisational costs. Of the 2 million individuals suffering from a work-related illness, 712,000 of these affected people were no longer in work, while of the remaining 1.3 million, 545,000 lost an estimated 19.5 million working days because of work-related illness (HSE 1995). According to this study, over half a million individuals in the UK had experienced stress or a 'stress-prescribed' condition, such as hypertension, heart disease or stroke (HSE 1995). In total, stress was the second most commonly reported condition (279,000 people), while a further 254,000 individuals described stress at work as causing or making their symptoms worse (HSE 1995). The incidence of stress is particularly worrying since it is linked to a wide range of illnesses and diseases. According to the HSE (2001a), prolonged or particularly intense stress can lead to heart disease, back pain, gastrointestinal disturbances, anxiety and depression (HSE/HELA 2001: 26). The HSE's definition of work-related stress is the 'adverse reaction people have to excessive pressures or other types of demands placed upon them'. These demands may include working hours, productivity targets and demands on emotional labour (Cooper 1985; Mackay and Cooper 1987; Schuler 1980; Deery, Iverson and Walsh 2000). By reviewing these key areas, we are more able to fully appreciate the considerable influence of workplace factors on OHS.

This chapter therefore introduces the day-to-day reality of workplace factors that relate to work organisation and the working environment, and their subsequent influence on OHS outcomes. We begin with a review of the impact of new technology on the working environment, before considering the factors that affect both the physical and emotional dimensions of contemporary work.

The working environment

> As society now recognises the importance of preserving our external environment from destruction, it is evident that our indoor environment, where most of us spend much of our lives, is just as important to our health, wellbeing and overall human and social development. We should therefore expect that our workplaces are healthy, environmentally friendly and safe.
>
> (Council of Civil Service Unions 1991, quoted in Baldry and Bain 1994)

Numerous studies suggest that members of the general public perceive the risks from outdoor air as being substantially higher than those from indoor air (London Health Education Authority 1997). However, there is evidence that indoor concentrations of many pollutants are often higher than those typically encountered outside (Jones 1999: 4536). Given that the majority of people living in developed countries spend much of their time indoors, concerns about indoor air pollution have steadily grown over the past twenty years or so. Indeed, one US study reports that 88 per cent of participants' days were spent indoors, with only 5 per cent of time actually spent outside (Robinson and Nelson 1995).

Since the 1980s, concerns over the health effects of indoor air quality have been stoked by the growing incidence of symptoms and illnesses attributed to indoor air pollution. In the 1980s and 1990s, numerous field studies on indoor air quality and 'sick building syndrome' were conducted, mostly in office environments. In 1982, the WHO recognised sick building syndrome (SBS) as occurring where 'a cluster of work-related symptoms of unknown cause are significantly more prevalent amongst the occupants of certain buildings in comparison with others' (Baldry and Bain 1994: 3). According to Wallace (1997), typical symptoms include:

- headaches and nausea;
- nasal congestion (runny/stuffy nose, sinus congestion, sneezing);
- chest congestion (wheezing, shortness of breath, chest tightness);
- eye problems (dry, itching, tearing, soreness, blurry vision, burning, problems with contact lenses);
- throat problems (soreness, hoarseness, dryness);
- fatigue (unusual tiredness, sleepiness or drowsiness); and
- dry skin.

Those afflicted experience individual, or combinations of, SBS symptoms on a daily basis, and the severity and incidence of the symptoms is related to the amount of time spent by sufferers in the affected work environment (Baldry and Bain 1994: 4). In 1992, the HSE estimated that between 30 and 50 per cent of new and remodelled buildings were affected by sick building syndrome, and within these buildings, 85 per cent of occupants were likely to suffer sick building symptoms. At a global level, the WHO has estimated that one-third of new or remodelled commercial buildings worldwide are 'sick' (Raw 1992). In the USA, it is estimated that SBS is a risk for up to 70 million workers (Baldry and Bain 1994). In addition, one Swedish study reports that between 600,000 and 900,000 people are exposed to an indoor climate that can affect their health negatively (Norlén and Andersson 1993), while a survey of all municipal workplaces in one Swedish city indicated that one-quarter had building-related health problems (Thörn 1998: 1307).

Origins of the problem

Some explanation for the prevalence of SBS is found in a range of cost-efficiency measures affecting building design and operation, which were introduced during the 1980s as thousands of new and remodelled buildings were used to house the growing population of white-collar workers. Since the 1960s, the number of white-collar jobs has massively increased in tandem with the expansion of public services (health, education and welfare). By the 1970s, white-collar workers accounted for around half of all those in employment in the major economies. Changes to building construction and operation over the same period are key contributory factors in SBS. As Jones (1999: 4536) explains, modifications in building design were driven by the need for increased energy efficiency, largely brought about by higher fuel costs since the 1970s oil crisis. In the 1980s, these cost pressures are likely to have been instrumental in the decision of ASHRAE (the body that sets air quality recommendations in the USA) to reduce the recommended air-refresher rates from 10 litres per person per second to 2.5 litres. This represented a 75 per cent reduction and significant savings for employers (and the environment) (Baldry and Bain 1994).

Other measures to improve energy efficiency included sealed units that improved insulation. This was accompanied by numerous modifications to the management of indoor environments (for example environmental control units that control temperature, humidity, lighting), while advances in construction led to a much greater use of synthetic building materials. Following these developments, it has been argued that buildings may now provide an environment in which airborne contaminants are readily produced and may build up to substantially higher concentrations than typically encountered outside (Jones 1999). One of the unanticipated outcomes of these innovations may be SBS.

Causes

No single factor has been attributed as the primary cause of SBS. The most common factors include inadequate ventilation and/or fresh air supply, high or low temperatures, low humidity, airborne pollutants (for example, formaldehyde, dust, microbiological contaminants). In addition, factors relating to work organisation are identified, such as job grade (that is, low job grades), lack of control over the physical environment (for example, the inability to open a window in sealed or 'tight' buildings) and gender, with females more likely to experience SBS symptoms. Some studies suggest that both gender and occupational grade have a marked effect on who is affected. In a Swedish study, Stenberg and Wall (1995) report an excess of symptoms amongst females (12 per cent compared to 4 per cent amongst males), the incidence of which was attributed to different hierarchical positions in the workplace.

Technological advances in air conditioning and environmental control systems have transformed building design in terms of dependence on fresh air. During the 1980s, systems that were capable of recirculating air became popular in buildings and aircraft. Although many factors have been found to contribute to SBS, commentators generally agree that inadequate ventilation and/or recirculated air are key factors (Raw 1992). For example, one prominent study reported that the proportion of recirculated air correlated with concentrations of contaminants in indoor air (Turiel *et al.* 1983). In addition, one German study of 8,000 building occupants found that sick building symptoms were more prevalent in air-conditioned buildings compared to 'fresh-air'/non-air-conditioned buildings (Kroling 1987, quoted in Raw 1992). However, Donnini, Van Hiep Nguyen and Haghighat (1990) report that while a greater proportion of fresh air will reduce carbon dioxide levels, formaldehyde and toluene, there may be a corresponding increase in dust, temperature and relative humidity. This illustrates the dilemma faced by many office workers in city-centre locations, where no respite from poor indoor air quality is provided by opening a window (assuming that it is not a sealed window unit), since traffic or other pollution will only exacerbate the problem. As such, workers in many buildings are at the mercy of heating, ventilation and air-conditioning (HVAC) systems. Moreover, there is little scope for individual manipulation of these conditions provided by this technology, which may be problematic since it is unlikely that individual workers will have exactly the same comfort zones in relation to temperature, drafts, lighting and so on. This may partially explain why lack of control over these conditions has been identified as a further factor in SBS.

To a large degree, building occupants' comfort is dependent on how well their employers operate and maintain the air-conditioning or HVAC systems. These systems have been shown to contribute to indoor air pollution in three main ways: low ventilation flow rates, incorrect installation and/or maintenance rendering the filters ineffective (for example, allowing circulation of pollutants around building spaces) and the production of pollutants from within the system itself (such as fungi and mould growth resulting from poor maintenance). Raw (1992) reports numerous cases of serious contamination in HVAC systems, with clear evidence of bacterial and fungal build-up in poorly maintained systems, while inadequate ventilation flow rates were identified as a causal factor of SBS symptoms in half of the 356 investigations conducted by the US National Institute for Occupational Safety and Health (NIOSH) between 1974 and 1985. The possibility that pollutants can circulate around building spaces (that is, the failure of air filters to trap all particulates) was demonstrated in a study conducted by the Society of Radiographers (1991). Their study attempted to discover the reasons for the high level of illness and symptoms reported by its members in a particular hospital. On analysing air quality in various parts of the hospital, a range of pollutants and particulates was found to circulate throughout the hospital, despite the presence of 'high-efficiency' air filters in the air-conditioning system.

The costs of poor indoor air quality

Based on Danish empirical research, Ole Fanger (2001: 149) argues that the quality of indoor air has a significant impact on the productivity of office workers. In comparing productivity levels before and after a pollutant was introduced, the author reports that the productivity of the subjects was found to be 6.5 per cent higher in good air quality conditions, and subjects also made fewer errors and experienced fewer SBS symptoms. These findings support earlier studies as reported by Raw (1992), which linked SBS to lower productivity. This is hardly surprising given the nature of the symptoms. How well would any of us perform at work if we were routinely suffering from headaches and lethargy, as well as a range of eye, nose and throat irritations?

The experience of these symptom clusters may also cause employees to be absent from work more regularly. However, the pressure to control costs has increased the emphasis on absence management in many organisations, pushing absence management programmes up the organisational agenda. For the occupants of so-called sick buildings, absence from work to allow recovery from SBS symptoms may simply not be an option. Absence management policies can be seen, therefore, potentially to conflict with any objective to prioritise employee health and well-being. Mind you, if that were a primary goal, the quality of the working environment would probably be managed quite differently.

Absence management

Developments in absence management in the UK are illustrative of the government's and employers' failure to link employee health with, amongst other things, workplace factors, along with a singular focus on short-term cost savings. In 1994, as a response to rising costs associated with employee absence, the UK government passed on the financial burden of absence costs to employers. The government believed that this action would motivate employers to improve the attendance records of workers, and so reduce costs. The government of that time might well be proud – a range of punitive absence management strategies are now employed in most UK organisations (including return-to-work interviews, home visits and various penalties for poor attendance, such as reduced holiday entitlement). One indicator of their success is found in various studies that find that a majority of UK employees are afraid to take time off work when they are ill (UNISON[1] 1998, TUC 1995).

A European Foundation for the Improvement of Living and Working Conditions (EFILWC) (1997) report has criticised this range of procedural measures (found in most countries) as simply an attempt to reduce absenteeism by tightening procedures without introducing any preventative measures. The report suggested that in attempting to reduce employee absence, employers should focus on acquiring safer equipment, climate control, task rotation, better information systems, work organisation and safety management. The benefits of implementing such measures can be found in Norway, for example, where a national absenteeism project found that the companies that had improved working conditions were also the most successful in reducing absenteeism (on average by 10 per cent a year) (*Health and Safety Bulletin* 262, October 1997). Based on these findings, a more successful approach to reducing absence and improving workplace OHS would take into account all aspects of the physical working environment as well as work processes. With that in mind, let us now consider the role of work processes in OHS in the context of working hours, productivity pressures and emotional labour.

Working hours and productivity pressures

Working hours have long been recognised as having a significant impact upon workers' health. In Sparks *et al.*'s (1997) meta-analysis[2] of the literature (covering the period 1965–96), the effects of working hours on health are examined, and a link between long working hours and ill-health is firmly established. Long hours of work, especially overtime, and vertiginous increases in work intensity for many workers are linked to the experience of fatigue. Fatigue has been linked to the three-fold increase in accident rates for individuals who work 16 or more hours (Sparks *et al.* 1997; see also Harris and Mackie 1972). In addition, worker fatigue has been attributed to a number of nuclear power plant safety compromises (Rosa 1995). Other interesting studies include that of Kogi, Ong and Cabantog (1989), who report on how 12-hour shifts in Singapore factories were abandoned because of their adverse effect on worker health, productivity and employee turnover. In addition, Äkerstedt (1994) reports how accident and injury rates trebled between 9 and 16 hours of work, while Laundry and Lees (1991) identified higher rates of major injuries in a yarn manufacturer during the latter part of 12-hour shifts (see also Lowden *et al.* 1999).

Workers appear to be all too aware of the relationship between working patterns and ill-health. In addition to the figures from the HSE quoted at the beginning of the chapter, one survey by the Manufacturing, Science and Finance Union (MSF) reports how workers made a direct connection between their experience of stress and sleeping problems to working hours and shift patterns (MSF 1997). This study also reviewed the extent to which employers were encouraging healthier living. Highlighting the low recognition given to employee health was the finding that few employers took action to restrict excessive working hours and only 21 per cent had carried out a health and safety audit. In addition, only half provided an occupational health service for employees. Compounding the direct effects of a long-hours culture are its associated lifestyle habits, such as inadequate diet, lack of exercise and heavy smoking, which in isolation can, of course, lead to health problems (see, for example, Maruyama, Kohno and Morimoto 1995; Sparks *et al.* 1997).

All of this goes some way to underline the potential value of the European Working Time Directive. The Directive formalises regular breaks during working hours and places maximum limits on working hours over 4-week periods. The magic number of 48 hours adopted by the Directive is ubiquitous in the literature, where it is shown that the incidence of heart disease and a wide range of mental and physical health problems rise exponentially when working hours exceed that figure. For example, Barton and Folkard (1993) report that employees[3] who worked 48 hours a week or more experienced a greater incidence of mental and physical health problems (for example, anxiety, cardiovascular problems, digestive problems and neuroticism) compared to those who worked fewer hours, while Buell and Breslow (1960)[4] reported that working more than 48 hours per week doubled one's chances of dying from heart disease. Despite the solid logic and support from considerable research evidence to support the 48-hour limitation, it is an accepted and increasingly common practice for employees to sign a 'voluntary' opt-out clause, thus removing any restrictions to their working time. As mentioned in Chapter Two, reasons for signing opt-outs may include financial hardship, threat of redundancy and limited promotion opportunities – all of which, we noted, highlighted the extent of workers' 'free choice' in accepting the associated health risks of working beyond the recommended limit on working hours.

The pressure on employees to accept unsafe working practices is highlighted by Cutler and James (1996) who cite the example of the 1988 Clapham rail disaster, in which thirty-five people were killed as a result of a wiring error. The UK Department of Transport (1989) recorded that a technician who had worked considerably long hours over an extended period during a rewiring programme at British Rail made the error. It emerged that the technician had worked for 13 weeks without a day off. Even after the disaster, the Rail, Maritime and Transport Workers Union (RMT) said that technicians continued to work an average of 57 hours a week. Following on from Nichols and Armstrong (1973), Cutler and James (1996) use this example to demonstrate the 'ambivalent relationship' between unsafe working practices and company policy. They argue that while dangerous working practices contradict formal safety policies, production pressures imposed by management place the worker somewhere between a rock and a hard place. Resultant errors can consequently be recorded as 'human error', since they contradict and deviate from formal company safety policies. Moreover, formal inquiry rulings of 'human errors' effectively absolve the organisation of blame for injuries and/or fatalities, ensuring that the bill for safety is dodged by employers and instead passed on to employees. Forms of payment include loss of life, injury, illness, and even prosecution.[5]

The demands placed on workers by intensive work regimes, productivity pressures and long working hours, are compounded by management's increasing invasion into workers' emotional territory. Over the past two decades emotion has been firmly placed on the organisational agenda. It is now widely recognised that 'organisations have feelings' (Albrow 1994, 1997); that they are sites of 'love, hatred and passion' (Fineman 1993) and the 'commercialisation of feeling' (Hochschild 1979, 1983) is a common occurrence. Indeed, as employers are openly engaging with 'hearts and minds' (Warhurst and Thompson 1998), it is indisputable that the management and manipulation of employees' feelings is securely tied to the idea of competitive advantage (Bolton and Boyd, 2003). The massive growth of interactive service work brings with it a focus on the delivery of high-quality customer service, often involving prolonged periods of interaction with sometimes unpleasant, ignorant and even aggressive people. While this might remind you of various work colleagues, our focus is on the rising trend of customer violence and the subsequent demands that this places on 'emotional labour'. Let us first review what is meant by 'emotional labour' before considering its potential impact on employee health.

Emotional labour

> We are maintaining a commitment to customer service which transcends merely being nice. It demands more than a plastic smile and 'have a nice day' . . . As well as using their brains, they are going to need to use their hearts and engage in what we call 'emotional labour'.
>
> (Dr Nick Georgiades, a former Director of Human Resources at British Airways, quoted in Blyton and Turnbull 1998: 69)

Hochschild (1983) offers an insight into the social actor's ability to work on emotion in order to present a socially desirable performance, as well as capitalism's appropriation of that skill. She emphasises how the self-management of emotion can, just as with physical labour, entail conscious effort and hard work (Bolton and Boyd, 2003). As Ritzer (1993) notes, society has come to expect McDonald's-style courtesy and friendliness in service workers. This requires workers to display emotions that comply with certain expression norms or rules of the organisation that help to create a desired 'state of mind' in the customer (Deery, Iverson and Walsh 2000: 2). In this context, employees are expected to 'appear happy, nice and glad to serve the customer in spite of any private misgivings or any different feelings they may have' (Erickson and Wharton 1997: 188). In addition to, and even fuelling, customer expectations is the level of competition within the service sector industries, which means that employers' efforts to shape workers' emotions and emotional displays have become an important focus of competitive strategy. As Bolton (2000: 581) notes, the introduction of a market rationality into the management of the British NHS has led to the term 'emotional labour' being used as part of a 'business model' of health care, where a nurse's caring skills are utilised as a resource. This 'commodification of emotional labour' (Hochschild 1983) extends to the service sector, where a degree of emotional investment is required to provide the 'friendly, cheerful and helpful' emotional display that we increasingly expect from interactive service workers. As Van Maanen and Kunda (1989) have argued, organisational culture management may simply mask managerial attempts to control not only what employees say, but what they feel. For Hochschild (1983: 186), emotional labour provides a new arena for the capitalist to exploit the worker: 'Capitalism has found a use for emotion management, and so it has organised it more efficiently and pushed it further.'

However, the consequences of constant and sustained demands for emotional labour on employee well-being may be considerable. As Wharton (1996) explains, the disjuncture between what employees might feel towards their customers and what they are expected to display may prove difficult to resolve and could cause considerable anxiety. This appears to be the case in Deery, Iverson and Walsh's (2000) case study of five Australian call centres, where they argue that excessive demands on emotional labour led to a higher propensity of stress, anxiety and emotional exhaustion amongst call centre agents. Similarly, Rutter and Fielding's (1988) study of prison officers found that demands on emotional labour were positively correlated to stress and negatively correlated to job satisfaction. A number of other 'dysfunctions' of emotional labour have been identified by various other commentators (Hochschild 1983; Cooper 1985; Maslach and Jackson 1985; Karasek and Theorell 1990; Morris and Feldman 1996). Given that over 76 per cent of employee jobs in the UK are provided by the service sector (Office for National Statistics, December 2000) – a figure that is likely to be comparable to, for example, US and Australian figures – the interaction between emotional labour and work-related illness, absence and low productivity could be considered an issue of some importance.

Because more research has been carried out since Hochschild's seminal work, we can now see a far more complex picture emerging whereby there is a crossover between Hochschild's private and commercial emotion management, the degree of effort made by individuals in conforming to organisational emotion rules as well as individuals' resistance to demands for emotion management (for a critique of Hochschild's analysis, see Bolton and Boyd, 2003). There are many examples of 'spaces of resistance and variance of effort' in the literature (for example, Taylor and Bain 1999; Bolton 2000). Sharon Bolton (2000), in her study of nurses, introduces the notion of 'gift exchange' where she argues that actors can do varying degrees of emotion work and that there is choice in what, when and how much and to whom they give.

Variance in how individuals deploy emotional labour is underlined by studies that record employees' resistance to scripted texts and behavioural rules (see, for example, Taylor 1998; Taylor and Bain 1999; Paules 1996). In addition to these are numerous examples of the coping mechanisms employed by emotion workers (for example, call centre agents, airline flight attendants). An amusing example is provided by one airline flight attendant, illustrating how humour is used as a coping mechanism. She describes an example of a fake performance review completed during a night flight, during which time the crew huddle in one of the galleys chatting while the passengers sleep:

> Cabin crew X always deals promptly with passenger requests. However, her manner is a little abrupt, particularly when she shouts in their faces to sit down and shut up. In terms of improvements, I have suggested that crew member X refrains from picking her nose and scratching her arse when walking through the cabin.
>
> (Bolton and Boyd, 2003)

This supports other research findings that show instances where humour in organisations is not only in the interests of 'group conformity' but can also act as a means of relieving boredom and 'letting off steam', and offering support and friendship (Collinson 1992; Fine 1998; Bolton 2001).

In another interesting study, Paules (1996) analysed the dynamics of resistance among waitresses. Where customer hostility or aggression was encountered, waitresses adopted a 'waitress as a soldier' persona whereby the customer's hostility was matched and even surpassed by the waitress. In doing so, Paules (1996) argues that these workers were able to maintain a sense of self-worth and deflected assaults on their dignity and adulthood. However, this coping mechanism may not be available to many other service workers, particularly those who experience high-level monitoring at work (for example, call centre agents). For those service workers who are encouraged to diffuse customer hostility with a 'customer is always right' approach, it is possible that in the process of their working day a raft of verbal abuse from vexatious customers may have to be absorbed. Worryingly, this approach may lead in turn to an acceptance that verbalised customer dissatisfaction, no matter how upsetting for the recipient, is 'just part of the job'.

One possible ramification of employee acquiescence is under-reporting. If patterns of under-reporting work-related injuries to the HSE provide any clue about the extent of under-reporting for verbal abuse, it is likely that the majority of incidents are not communicated to management.[6] While there is no legal requirement to report verbal abuse, an already alarming picture is emerging from current figures for the number of reported incidents involving physical violence.[7] For the period 1999/2000, 5,034 minor incidents (over 3 days of absence) and 686 major incidents were reported (HSE 2001a).[8] Mayhew (2002: 22) notes a range of other reasons why workers may choose not to report verbal abuse, such as embarrassment and the influence of organisational culture.

Based on an assumption that service workers comply with employer rules on emotional displays and behaviour, the growing incidence of customer violence may increase both the volume and intensity of 'emotion work' in the international service industries. We now go on to consider customer violence in some detail, using case-study evidence from the UK rail and airline industries to demonstrate the linkages between demands on emotional labour and customer verbal and physical abuse.

Violence at work

> Violence to staff at work from members of the public is at an unacceptable level ... no-one should have to accept verbal or physical abuse as part of their job.
>
> (Bill Callaghan, Chair of the Health and Safety Commission,
> 8 June 2000)

Violence at work is a widespread but largely unrecognised problem across all sectors (Chappell and Di Martino 1998). Recent headline events, such as the flight attendant attacked with a broken bottle leaving her scarred for life, and the tripling of assaults on railway workers during the period 1995–8, highlight the growing problem. A TUC survey outlines the increase in verbal abuse and physical violence experienced by UK workers, with one in five workers subject to a violent attack or abuse at work every year (TUC 1999b). Moreover, the report finds that women, especially those aged between 25 and 34, are most at risk. This study follows the HSE's 1995 study of self-reported working conditions, which found that 1.3 million people reported being physically attacked by a member of the public. Half of these had been attacked between one and four times in the previous year and a further 10 per cent had been attacked at least five times in the same period. The greater propensity for women to experience customer violence compared to men was also identified by this study (8 per cent of women compared to 6 per cent of men). In addition, the 1997 British Crime Survey reported 523,000 physical assaults by members of the public against workers. In total, almost 650,000 workers had experienced at least one violent incident during the year (HSE 1999b). The HSE also report that around one quarter of a million assaults at work resulted in some type of injury, from bruising to broken bones. The majority of victims were upset by their experience, suffering a range of symptoms such as anger, fear and difficulty sleeping (HSE 1999b).

While attention has traditionally focused on physical violence in the workplace, more recent evidence points to the impact and harm caused by non-physical violence, such as verbal aggression or abuse. In a 1994 survey by the Canadian Union of Public Employees (CUPE), verbal aggression was stated to be the leading form of violence against employees (Pizzino 1994). Mayhew and Quinlan (1999) report from their studies of a number of occupational groups in Australia that verbal abuse (and threats of physical assault) were a common phenomenon among taxi drivers, fast-food workers, hospitality and bar workers, nursing and hospital staff and teachers. Interestingly, females doing the same job tasks as males were often more frequently verbally abused than their male counterparts (Mayhew and Quinlan 1999: 19). The prevalence of non-physical violence in the UK is highlighted by previously mentioned British Crime Survey figures for 1997, where 703,000 cases of threats by members of the public against workers were recorded (HSE 1999b).

Customer violence, in the forms of both verbal and physical abuse, is particularly prevalent in the UK's railway and airline industries. While most interactive service workers have only short-term contact with customers, railway and airline staff are relatively unique in that they may experience long-term contact with abusive passengers during a long train or aircraft journey. In these cases, staff are effectively 'hostages' of abusive and/or disruptive customers in aircraft or train carriages, and consequently the actual strain in dealing with these situations, never mind the risk to health and safety that they present, may be more pronounced. In the confined environment of a crowded aircraft at 30,000 feet, cabin crew cannot simply walk away from threatening or violent situations. Nor is there the prospect of calling for immediate outside help. The nearest police assistance might be hours of flying time away. Railway employees face similar problems in that staff may be working alone in carriages or other areas where they have no immediate access to, or contact with, colleagues.

The HSE's Railway Industry Advisory Committee (RIAC) (1999) reported that injuries to railway employees as a result of passenger violence were up by 40 per cent compared to three years previously. In the three years ending on 31 March 1999 there were nearly 1,000 injuries suffered by railway staff that were of a sufficiently serious nature to be reported to the HSE (RIAC 1999). One report found that certain groups of railway employees were more likely to suffer an assault, which was related to the level of contact they had with customers. As Harris (1999) reports, train operators were least at risk of assault by a customer, while revenue-control inspectors were most at risk.

Similar trends of customer violence are apparent in the international airline industry. The International Transport Workers' Federation (ITF) has documented reports of cabin crew members being punched, head-butted, kicked in the back, bitten on the cheek, throttled, hit by a bottle, and, in the case of a KLM cabin crew member, stabbed. Cabin crew in one airline reported around 100 incidents of verbal or physical abuse per month (*ITF News*, June 1998). In one UK study of airline cabin crews, 70 per cent of respondents reported an increase in the number of abusive passengers in the past year, while the number of symptoms suffered by respondents (including eye/nose/throat irritations, fatigue, headaches) positively correlated with the experience of abusive passengers (Boyd and Bain 1999). In addition, 32 per cent of respondents considered abusive passengers as a high OHS concern, with 21 per cent of respondents complaining about abusive passengers 'often' or on 'every flight'. Despite the presence of this serious occupational hazard, less than half of all respondents had received in-depth training on dealing with violence (Boyd and Bain 1999).

The growing trend of abusive passengers is confirmed by Civil Aviation Authority (CAA) figures, which show an increase from 1,205 incidents to 1,250 reported incidents over the periods April 1999 to March 2000 and April 2000 to March 2001 respectively. For the latter period, some 78 per cent of all incidents involved male passengers, about two-thirds of whom were in their twenties and thirties, with 95 per cent of incidents occurring in economy-class seating (Department for Transport, Local Government and the Regions 2001). In eighteen of the 1,250 incidents, a passenger had to be physically restrained by handcuffs (up from thirteen in 1999/2000). The most common passenger misbehaviour, as for the previous year, was smoking in the aircraft's toilets, where in sixty-three separate incidents, the passenger had either started a fire or disconnected the smoke detector. Verbal abuse either to cabin crew or other passengers accounted for 44 per cent of reported incidents over issues such as dissatisfaction with the level of service or arguments over reclining a seat, while alcohol was identified as a contributory cause of disruptive behaviour in almost half of all incidents (DTLR 2001). Amidst the media attention on 'air rage' and passenger violence, CAA is quick to point out the relatively low frequency of these incidents, noting that for the period covered by the above figures UK airlines carried 104 million passengers on 1.1 million flights, making the incidence of *reported* disruptive passenger incidents about one in 880 flights.

The figures also show that on only thirteen occasions (up from eight in 1999/2000) out of the total 1,250 reported incidents was a diversion to the nearest airport necessary. This provides some indication of how successful cabin crews are in managing these situations. Their adeptness in diffusing violent situations may, however, remove (or at least subdue) the mood of urgency for implementing remedial measures. Their position as emotional punch bags is highlighted by one cabin crew respondent:

> Over the years the airline industry has taught its cabin crews to be very subservient towards passengers. The passenger is 'always right'. The customer is fully aware of this and takes full advantage of the situation. They know they can say anything they want to cabin staff and get away with it, and they usually do. I have been employed as a cabin crew member for the past twenty-one years and I have had to suffer a range of indignant remarks and affront on a daily basis.
>
> (Quoted in Boyd 2002: 158)

In a study of three UK airlines and one UK railway company, Boyd (2002) reports a high prevalence of customer violence. Almost three-quarters of respondents experienced verbal abuse from customers at least once a month in a myriad of forms, including sarcasm, condescending remarks and swearing. Overall, more than half of respondents reported an increase in the experience of verbal abuse from passengers in the previous year (53 per cent). A higher proportion of railway employees reported an increase (61 per cent) compared to airline employees (50 per cent). Almost three-quarters of all respondents reported that they experienced verbal abuse from passengers at least once a month (74 per cent).

In addition, 60 per cent of all respondents had experienced at least one type of physical abuse, while 37 per cent had experienced at least two types of physical abuse and 26 per cent had experienced at least three types of physical abuse, including being punched, kicked, slapped and being spat at. Over half of airline respondents (55 per cent) had experienced one type of physical abuse, a figure that (in a single study) is markedly higher than the overall official CAA figures. CAA statistics show that in 1998, only ten incidents involving acts of physical violence were reported, dropping to five in 1999 with a slight increase in 2000 to six incidents. In each of these incidents the passenger was restrained and police met the aircraft on landing, suggesting that it is only the most serious incidents that are reported to, and recorded by, the authorities. The more regular events such as those described below by respondents appear to be ignored in official statistics.

> A drunk passenger kissed and slavered down the back of my neck while I was serving from the trolley. I've also suffered bruising from being kicked and a man poked me with his walking stick.
>
> (Quoted in Boyd 2002: 160)

> I have been grabbed by my testicles. I have also had a drink thrown over me. I've been spat at and pushed over.
>
> (Quoted in Boyd 2002: 160)

> I was attacked on the train when I was on duty. The man who attacked me was arrested but I haven't slept for six months since and I am afraid to go to work but I never speak about it.
>
> (Quoted in Boyd 2002: 161)

The demands made on respondents' emotional labour are further underlined by the following remarks:

> Myself and close colleagues have not been involved in any physically abusive situations, but plenty of verbal . . . almost every day, we meet with aggression due mainly I feel, to stress. We are the frontline people of the airline and therefore a target for every petty problem that passengers are experiencing, whether in relation to their flight, family or business. Things get magnified out of proportion so often, and even after you have tried to placate the person, they never back down and apologise for their behaviour which leads to bad feeling on all sides. I feel

now I am on the defensive whenever someone approaches me by assuming that they are going to have a go at me.

(Quoted in Boyd 2002: 161)

I cannot believe people today. I've had one passenger urinate at my window right in front of me. I've had a man threaten to get me after work. There was also a woman claiming she'd put a curse on my unborn child while I was heavily pregnant.

(Quoted in Boyd 2002: 161)

The last letter from the company said something around the line of no matter what the situation is, we always need to look as though we are in control of the situation because the coach may be full of passengers. But the problem is, what if six drunk lads were causing trouble by annoying or verbally abusing other passengers? How would two female members of staff sort that out and look as though they were in control of the situation?

(Quoted in Boyd 2002: 161)

From the above comments, it is clear that customer violence causes a degree of emotional upset to recipients. Boyd (2002) reports that the experience of verbal abuse made almost one-quarter of respondents feel intimidated and almost half feel angry. In addition, well over half of respondents reported reduced job satisfaction as a result of experiencing verbal abuse. Common additional remarks from respondents about how the experience of verbal abuse made them feel included in the questionnaire were feelings of 'worthlessness', 'depression' and 'increased stress levels'. According to one respondent,

A combination of emotions is experienced and sometimes you feel like this so often you stop noticing how you feel.

(Quoted in Boyd 2002: 161)

The above comment alludes to notions of 'emotional numbness' and alienation from one's self as described by Hochschild (1983). The process of absorbing, diffusing and managing constant streams of verbal abuse appears to be stressful and wearing, suggesting more than a tenuous linkage between the demands made on emotional labour in the workplace and workers' health and well-being. It could also be argued that the regularity of these incidents introduces the dimensions of 'volume' and 'intensity' to emotional labour – which previously have only been applied to physical labour. However, a particularly worrying aspect of the study, and one that reflects wider trends in OHS, was the level of under-reporting of incidents involving verbal abuse. Boyd (2002: 158) reports that a high majority of respondents simply accepted verbal abuse as 'part of the job'. Moreover, a high majority of respondents thought that 'it was a waste of time' reporting these incidents since management was not interested or management might think that they were not 'up to the job'.

In all of this, a range of profit-centred strategies can be seen to perpetuate customer violence, where they create fertile conditions for both physical and verbal abuse towards staff. For example, Boyd (2002) identified the main reasons for customer violence in the case-study companies as alcohol consumption, delays, internal environmental conditions inside aircraft and train carriages, disputes over baggage and the failure of the companies to meet customers' expectations. As respondents explain,

> Everything seems to revolve around profit, for example, we sell a lot of beers on Saturdays.
>
> (Quoted in Boyd 2002: 163)

> Airlines increasingly use inexperienced, younger staff as ground staff for cost reasons – lack of experience/maturity/age often means overbearing passengers or those under the influence are still allowed to board/travel when perhaps stricter handling would deny them boarding. Airlines should invest more in experience of all staff in all areas and encourage consistency of strict rules in application.
>
> (Quoted in Boyd 2002: 163–4)

> Only now is the company backing up the crew on this issue. In the past it was discouraged to take any action on violent or abusive passengers for fear of bad publicity or non-repeat custom for the company.
>
> (Quoted in Boyd 2002: 164)

Also highlighting the low priority given to employee health and safety is the apparent lack of safety equipment and safety training provided to staff. As railway respondents explain,

> The present security measures are pretty useless: CCTV – old, fuzzy pictures and cameras are too far to be of use. Panic buttons still mean it takes the police at least 3 or 4 minutes to respond, sometimes as long as ten minutes.
>
> (Quoted in Boyd 2002: 164)

> Mobile phones are supposed to be on every train but in practice they're not on unless we have a manager on board. We are less likely to have them at the weekend when there are fewer staff and you probably need it more for safety reasons because there are less of you. Also the lack of a phone can indirectly lead to abuse because if you're late you don't get all the information about connections for example. To call the office you've got to wait in queues at public phones. There should be a staff phone on

every train. At the moment we have fixed staff phones but only about 40 per cent of these are working.

(Quoted in Boyd 2002: 164)

I have never received counselling or further training as a result of passenger verbal or physical abuse, which isn't good enough.

(Quoted in Boyd 2002: 164)

There are no control measures such as personal alarms or CCTV on the trains. You could be on another planet. 'Air rage' has been going on for years on trains.

(Quoted in Boyd 2002: 164)

This range of cost-minimisation policies operates alongside the abilities and goodwill of staff, who, in conscientiously attempting to defuse potentially violent situations (with or without the benefit of targeted training), may be unwittingly collaborating in their further exploitation. For example, by deploying their emotional labour as a 'band-aid' for management's failure to address and remove the triggers or sources of customer violence, these employees may be removing the urgency for intervention at both an organisational and a government level. At the same time, the popular provision of 'conflict management' training does nothing to remove the causes of customer violence (for example, the sale of alcohol), and instead manages only the symptoms. Similar arguments could be made about 'stress-management initiatives', all of which illustrate the creativity of management in making workers adapt to their construction of the workplace and the demands and strains that prevail within it.

The above case studies of customer violence pull together some of the macro- and micro-level factors influencing OHS policy, practice and outcomes discussed so far. For example, a self-regulatory approach is shown to provide employers with considerable freedom to determine what is an acceptable risk to employees in dealing with violent customers, while the porous language contained within health and safety legislation (for example 'reasonably practicable') appears to provide employers with enough excuses not to remove the triggers of customer violence and suitable justification to simply treat the symptoms (to varying degrees of effectiveness). Also at a macro level, the role of the trade unions in both industries has been instrumental in raising the profile of this serious health and safety risk, and it is this that has led to the introduction of new initiatives, such as 'Managing Violence at Work: A Partnership Approach', launched in June 2000. By now, we are fairly suspicious of the term 'partnership' within the present regulatory framework, but national recognition of an important OHS issue is, nonetheless, a step forward.

At a micro- or workplace level, we can see the influences of management style, organisational culture, work organisation and work processes affecting the occurrence and severity of customer violence. Crucially, a range of policies and working practices appears to exacerbate and even perpetuate customer violence, while the evidence of inadequate training for staff may mean that in many cases workers are left to their own devices to deal with stressful, difficult and dangerous working conditions. In terms of the bigger picture of occupational violence, Bowie (2002) provides an interesting four-way typology of violence at work that organises the macro- and micro-level factors mentioned above. The four types are listed as 'consumer-related violence', 'relationship violence' (for example bullying at work), 'intrusive violence' (criminal intent by strangers), and 'organisational violence' (for example the creation of an oppressive and violent workplace that may provide the triggers for violence from staff or clients). Bowie (2002: 11) explains organisational violence in more detail: 'Many earlier responses to workplace violence by employers deflected attention from a potential key contributor to workplace violence – the ways organisations are structured and managed.'

In relation to the railway and airline survey findings, we can clearly see consumer-related and organisational violence. The survey would also suggest that the legislative environment is related to the notion of organisational violence where policymakers can create fertile conditions for the perpetuation and exacerbation of violence.

Conclusions

This chapter has presented another piece of the OHS jigsaw. From the evidence presented, we can observe the knock-on effect of various political, economic and technological influences on how work is organised and the quality of the working environment. The domino effect appears to extend to employee health, with a range of work-related injuries and illnesses displaying clear linkages to a range of policies covering work organisation and the working environment. The theme of work intensification resonates from this chapter, whereby demands on both the physical and the emotional dimensions of work appeared to be growing. This was evidenced in part by the continuance of a long-hours culture, productivity pressures and service-quality demands. In pursuit of productivity gains, it appears that little acknowledgement has been given to the swaths of literature on the detrimental impacts of all of the above. Questions might be raised as to whether this show of ignorance is the result of a straightforward underestimation by management of the interaction between work organisation and employee health, or whether it is the case that workers are being knowingly and deliberately exploited. Equally, is it more the case that prevailing economic conditions constrain employers' ability to honour policy statements on OHS? We can consider some of these positions in the forthcoming industry case-study chapters, while applying our framework of macro- and micro-level factors, which will allow us to unravel the complex and often contradictory portrait of HRM and OHS in practice.

Part II

HRM and OHS in practice

5 HRM and OHS in the international airline industry

> In an industry like ours, where there are no production lines, people are our most important asset and everything depends on how they work as part of a team. This means that, to get the best results, managers have to care about how they (the employees) live and function, not just about how they work and produce.
>
> (Sir Colin Marshall, ex-chairman of British Airways, quoted in the *Financial Times*, 1984)

The airline industry is a popular playground for HRM policies and practices and so provides a cornucopia of insight, both in terms of the location of OHS in HRM agendas and in terms of the complex array of factors that influence OHS policy, practice and outcomes. These include an intensely competitive market environment, a relatively loose regulatory framework and a self-regulatory approach. In an industry that claims 'people are its most important asset', one might expect 'good practice' in terms of OHS. Equally, we might expect modern aircraft to offer a salubrious working environment, while the safety-sensitive nature of the industry might lead us to believe that the highest quality OHS standards are continuously met.

While corporate statements certainly appear to support and reinforce such expectations, a growing body of international research provides persuasive contradictory evidence suggesting that quite the opposite is true. This creates a series of conflicts and contradictions in relation to the principles of HRM and our expectations of health and safety management. First, there is some evidence to suggest that airlines are fully aware of a range of health and safety risks in the aircraft cabin, yet do little to minimise or remove these because of the costs involved in taking remedial action. If this were the case, it would not only ridicule their pious mission statements, but would also suggest that in one of the world's most safety-sensitive industries, worker health and safety is routinely exploited in the pursuit of higher profits. Second, in an industry that depends so heavily on cabin crews' commitment to cultural values such as service quality, the diminution of employee health would run counter to service and safety goals, as well as potentially undermining employee commitment. Such conflicts and contradictions fit well with the 'rhetoric and reality' debate in HRM. This theme is dominant in our exploration of airline cabin crews' working conditions. This chapter identifies a range of conflicts and contradictions in the international airline industry's HRM policies and practices in relation to their espoused goals, values and objectives, for example, the pursuit of superior service quality and safety standards. In explaining these anomalies, we draw on the wide range of macro- and micro-level factors covered in the previous chapters.

Our journey begins with a review of the competitive climate in the international airline industry, which is followed by a discussion of aviation regulatory frameworks and agencies, where some of the implications and consequences of relatively unrestrained competitive practice in the international airline industry are outlined. The chapter then goes on to discuss specific OHS issues and outcomes in relation to the aircraft working environment and cabin crew working practices. In doing so, we examine a wide range of primary and secondary research material in which the reality of cabin crew work is shown to clash quite spectacularly with airlines' professed approach to people management.

Competition and survival in the airline industry

Since the 1978 US Airline Deregulation Act, civil aviation has experienced profound changes, most notably changes to ownership, market liberalisation, and the emergence of global 'mega carriers' and global alliances (Blyton *et al*. 1998). Following deregulation in the USA and Europe, there has been a significant shift in the regulation of the industry, entailing a reduction in the extent of state involvement and its replacement by greater market regulation. The drive to secure mergers and alliances has occurred on a global basis, while the increased competition from new-start 'low-cost' airlines has provided an added 'bite' to market forces.

Waves of restructuring within the airline industry have followed deregulation, where a core objective has been to create a seamless transportation system that provides a clear service concept to consumers worldwide. Alliances between airlines have led to the sharing of computer reservations systems and international code sharing, allowing customers to travel seamlessly (on a good day!) through several continents on different carriers using just one set of tickets and checking in baggage at the first departure point. There are some 400 alliances involving more than 170 carriers – an exponential rise from just ten in 1983 (*Guardian*, 22 September 1998). Evolving from restructuring are three mega groups, namely the Star Alliance, Oneworld and Qualiflyer groups. These major alliances account for around 46 per cent of the world market, with the largest alliance serving 578 cities in 108 countries with 1,334 aircraft and 210,000 employees (Blyton *et al*. 1998).

Consolidation has not been the only survival strategy employed. Worldwide, the most visible effect of the relaxation of state regulation of airline markets has been the growth in the number of independent low-cost airlines. The new entrants have typically been able to offer low-cost flights on the basis of lower overheads, by operating from second-tier airports, engaging in direct selling, and by offering a very basic service (for example, no catering). In response to the competitive challenge presented by the new entrants, a number of incumbent airlines have sought to gain a presence in the low-cost market by creating their own low-cost airlines. In 2002, Qantas announced plans for a low-cost international carrier, aiming to 'take off' in late 2002, which is based on the model used by UK and Irish budget carriers such as Easyjet and Ryanair (*Guardian*, 3 April 2002). Another example is British Airways (BA), which set up the airline Go in 1998 to compete with Ryanair and Easyjet under a flurry of magniloquent press releases and marketing campaigns. Go, however, was soon to be halted by one of its main rivals, Easyjet, when the victor bought over the vanquished airline in May 2002. The irony of the buy-over is notable given the 1998 court action by Easyjet's founder, Stelios Haji-Ioannou, to block Go's establishment, where he claimed BA was subsidising the airline illegally. After losing the case, Easyjet ran a competition asking customers to guess the extent of Go's losses (*Guardian*, 4 May 2002). The new combined airline of Easyjet and Go (a steal at £374 million after being sold by BA for £100 million in 2001 to a management team backed by the venture-capital firm 3i) will employ some 2,800 staff, operate fifty-seven aircraft, and carry more passengers than Ireland's Ryanair, the present low-cost leader in Europe.

Highlighting the increasing market share of low-cost airlines are forecasts that low-cost airlines will take a 25 per cent share of the European market by 2010 compared to 5 per cent in 2002 (Financial Times Information Limited 2002). This starkly contrasts the vicissitudes of incumbent airlines. BA, for example, has axed 13,000 jobs since the September 11 terrorist attack in the USA, and by March 2004 the total job losses since 2000 will be around 20,000 (*Guardian*, 14 February 2002). The US airline industry has also been reeling following the terrorist attack and Congress has since provided a $15 billion bailout for the industry (*Guardian*, 18 May 2002). All of these developments have taken place at the same time as the liberalisation and harmonisation of the industry's regulatory framework.

The regulatory environment

International civil aviation has operated since the Second World War on the basis that all states sign the Chicago Convention of 1944, thus joining the International Civil Aviation Organisation (ICAO). The scope of the ICAO is to 'agree principles and arrangements' that transpose into non-binding resolutions. The ICAO framework is supplemented by European collaboration in a number of forums, in particular, the European Civil Aviation Conference

(ECAC) and the Joint Aviation Authorities (JAA). ECAC is a pan-European body, covering thirty-eight countries, providing a forum for policy formulation across a range of civil-aviation issues. The JAA is an informal grouping of National Aviation Authorities from (currently) thirty-three states, which since 1970 has developed aviation standards known as Joint Aviation Requirements (JARs) with the purpose of securing common safety standards across Europe. These standards have generally been adopted into Community law through EC Regulation 3922/91. In the UK, JARs[1] provide quite different requirements on, for example, air crew working hours and physical working conditions, to those laid out in the Working Time Regulations (1998) and the Management of Health and Safety at Work Regulations (1992). While the latter regulations technically apply when aircraft and air crews are on the ground, once out of UK airspace these becomes obsolete.

Across the international airline industry, there is what can only be described as a complete void of regulations relating to basic employee health and safety issues such as rest breaks, hygiene, air quality, ergonomics and rest facilities for employees. While the ICAO requires member states to set limits on flight and duty times, it does not regulate the airlines' operating practices or specify recommended limits, so conditions vary from country to country. In addition, JARs provide only 'recommended' levels for air quality pollutants such as carbon dioxide and ozone. And, unlike most other workplaces, the aircraft cabin is not routinely subject to external inspections, meaning that it is primarily left in the hands of individual airlines to ensure these 'recommended' standards are maintained on flights.

The gaps in health and safety regulations have been recently highlighted by the Flight Attendants and Related Services (NZ) Association (FARSA), which represents some 1,500 flight attendants in New Zealand. Their recent submission to the New Zealand Select Committee, in relation to the Health and Safety in Employment Amendment Bill, challenges its exclusion of air crew, whom they argue are exposed to a range of health and safety risks – many of which we cover later in this chapter. In addition, the trade unions argue that the NZ Civil Aviation Authority is not the appropriate body to manage air crew health and safety issues because its interests lie more in the safety of aircraft and passengers. As is the case in the USA and UK, the NZ CAA[2] regulations currently contain no measures intended to protect the health and comfort of flight attendants (FARSA April 2002, Issue 89: 18) – an astonishing omission considering the safety-sensitive role of these individuals.

While cabin crews fulfil the function of physical pieces of safety equipment, it appears that on an international scale there is little in the way of interventions that ensure their comfort and well-being in the workplace – factors that relate directly to flight attendants' ability to perform at an optimum level during emergency situations. It is also clear that crash survivability is the singular concern of the aviation regulatory bodies. The UK's CAA openly comments that it is not concerned with issues such as rest breaks or ergonomic considerations for cabin crews (CAA 1999). The hard issues of training, familiarity with safety and emergency equipment and knowledge of safety and emergency procedures remain the primary foci of the CAA and JAA, despite an abundance of international research evidence that describes a quite frightening range of health hazards in aircraft cabins.[3] The observed apathy of the aviation bodies suggests that factors such as poor air quality, shiftwork, manual handling and workloads are not recognised as potential health and safety risks. The well-known detrimental effects of these risk factors on health and well-being must, therefore, be neutralised at 30,000 feet – a clearly ridiculous explanation, but one that fits with the indifference shown by international governments, regulatory bodies and airline companies.

Progressive relaxation of state regulation and greater competitive pressures have given greater emphasis to making higher profits as well as finding ways to reduce costs. While the aircraft industry vigorously argues that safety is never compromised by commercial considerations, the reality is that safety involves significant operational costs, such as those relating to the thoroughness and efficiency of aircraft maintenance checks and employee training. The fact of deregulation and liberalisation in the industry is that airlines now have relative freedom to 'cost' safety within the prevailing environment of increasing competitive and economic pressures. And that appears to be exactly what is happening. Already industry representatives are openly demanding that internal harmonisation of safety regulations between Europe and the USA should follow the 'least cost-safe option' that is, where the regulations differ, the cheapest ones should be automatically applied (ITF 1998a). As the International Transport Workers' Federation (ITF) has argued, the onus appears to be more on ensuring that airlines can operate flexibly and without cost duplication in a global market than on the need for a rational set of international aviation rules operating to the highest safety standards.

The precedence of cost over safety is also illustrated by the lack of action taken by airlines and regulatory bodies following an £18 million investigation by the US National Transportation Safety Board (NTSB), which traced a number of explosions on aircraft to the positioning of air-conditioning units directly beneath the central fuel tanks in Boeing 747 aircraft. The explosions occurred when heat from the air conditioning units produced fuel vapour when the aircraft was stationary, which then ignited the central fuel tank. According to the NTSB, 'many hundreds of thousands of planes could be at risk' (*Herald*, 7 June 1997). Regulatory bodies have been slow to act, which is hard to comprehend unless we consider the financial implications of remedial action. As in the case of other remedial measures that would remove significant OHS risks (for example, increasing seat pitch to provide more legroom to reduce the risk of deep vein thrombosis), addressing this risk would present a massive cost to airlines, not only from the design costs, but also the lost revenue from grounding aircraft.

The regulatory bodies' apparent accommodation of airlines' business interests points to the possible adoption of a 'dual mission' that involves being sensitive to business concerns while at the same time attempting to uphold industry safety standards. An inherent danger of 'dual missions' is the subordination of safety oversight and regulation in favour of perceived short-term economic advantages and efficiency for operators. A tragic example of this is the Valujet crash in Florida in May 1996 when 110 people were killed. From its date of conception in October 1993 to the day of the crash, the US government publicly and regularly lauded Valujet as an example and inspiration to the industry. In the aftermath of the crash, it emerged that the Federation Aviation Administration (FAA) had conducted twenty-one separate investigations into serious safety incidents on Valujet flights prior to the crash, and had faulted the company for forty-three incidents in a September 1995 report. In addition, the FAA investigation into the crash reported the aircraft as being 'unfit to fly'. An inspiration and example indeed, that is, for greater regulation and enforcement action.

The regulatory frameworks in the international airline industry have received considerable criticism in the wake of restructuring and new employment practices. In particular, the emergence of 'virtual airlines' has raised concerns over the accountability of airlines and companies. The term describes a company that offers air travel from one place to another under a global brand, but whose services are supplied by contractors, franchisees and alliance partners. According to a leading industry analyst,

> In an ideal world, what's an airline going to be? As they move to becoming virtual airlines, they're going to want to rent aircraft by the hour and turn them into infinitely variable costs. Airlines will be a brand, a yield management system and access to a distribution mechanism . . . they don't need to own anything.
> (Kleinwort Benson, quoted in *Air Transport World*, May 1997)

Keeping control though a complex web of contractual agreements is not just a problem for the company, but for the regulator. A regulatory nightmare is created when aircraft are owned by one company in any given country, maintained by someone else and possibly flown by crews in a different country. However, one of the most serious threats to regulatory control over civil aviation is the risk of airlines stepping outside proper regulatory oversight altogether. The term 'flag of convenience' was coined by the ITF to refer to the maritime industry where ships are registered in countries with lower safety standards than the country in which the ship will operate. In Europe there is already a problem of charter airlines using one country as an operational base while using aircraft registered in a different country. In these cases, the national aviation authority of the country from which the aircraft operates has no responsibility for the safety standards of the aircraft. Regulatory loopholes allow UK tour operators, flight bookers and 'seat only' companies to use non-EU 'flag of convenience' aircraft and crews. However, following a crash in February 1996, when 176 German tourists were killed returning from the Dominican Republic in a Turkish Boeing 757, the largest German tour operator issued a list of approved foreign airlines to charter carriers, and banned sub-chartering outside the list (*Observer*, 19 May 1996). Asia appears to be a particular honeypot for safety violations. In August 1998, Grand Air was grounded for flying routes with unqualified pilots, while in September 1998, Asian Spirit was grounded for a 'non-compliant attitude to safety'. Indeed, since deregulation in 1994, four out of five start-up carriers in Asia have spent time grounded by the authorities because of safety violations. While the international maritime industry appears ready to live with the fact that one bulk carrier goes down per week (ITF 1998a), similarities in the use of flag of convenience carriers and the prioritisation of cost over safety in the international airline industry suggests that there is every chance that it will follow suit. All of this suggests that a 'low-cost' safety approach is routinely adopted in the international airline industry, partly because of competitive economic conditions, but also because of regulatory loopholes that are big enough for aircraft to fly through. Let us now consider the impact of these factors on employment practices.

Employment practices

The Board of British Airways believes that excellent standards of safety and health are essential for the well being of our people, reputation and business performance. We are committed to ensuring all our activities worldwide are conducted in accordance with industry best practice.

(Robert Ayling, Chief Executive of British Airways, excerpt from BA's Corporate safety-management policy)

British Airways is, without doubt, a key trend-setter in the international airline industry. Consequently, industry 'best practice' will often be bench-marked against BA's espoused policies and practices. If we believe in the content of the above policy statement, then this should not present a worry. However, a review of BA's and other airlines' policies on subcontracting, employee pay and working conditions, and the cabin working environment, suggests that the authors of the policy statement have, to a large degree, lost touch with reality.

The developments in labour market restructuring that were discussed in an earlier chapter are mirrored in the international airline industry. Amongst European airlines, BA took an early lead in outsourcing non-core activities such as catering and vehicle management and maintenance (European Commission 1997: 202–3), while the outsourcing of some ticketing services to India began back in 1991. The global trend of subcontracting is supported by the Blyton *et al.* (1998) study, in which more than half of the unions questioned reported a global trend of subcontracting services, most notably in aircraft maintenance. This is explained in part by the cost-intensive and increasingly specialised nature of maintenance. In addition, newer aircraft require less frequent maintenance, making it harder for individual airlines to justify the expense of in-house maintenance. Already, the contracting-out of aircraft maintenance has led to massive job losses and the transfer of work and workers into companies where conditions and benefits for employees are often inferior to those previously enjoyed. As maintenance becomes an increasingly global affair, unions are concerned that airlines will have increased opportunities to relocate work around the world, with the aim of chasing lower employment costs.

These fears appear to be well founded, based on the significant movement of customer service jobs (notably to non-union sites) to date. In July 1999, BA announced plans to shut down its telephone-sales operations in New York with the loss of around 600 jobs. According to the ITF, many of these jobs were moved to a non-union site in Florida, while other work would go to India. A BA spokesperson said, 'BA has decided to outsource telephone sales to lower costs' (*ITF News*, April 1998). Non-union sites also appeared to be favoured in BA's Indian operations, illustrated by the expanded operations of WNS – a non-union subsidiary company. In October 1998, the British Airways Employees' Union in Bombay wrote to their local BA management to complain that 'the company has been using this subsidiary (WNS) to create a workforce parallel to the unionised workforce of BA with the intention to destroy the organised and collective strength of the workmen of BA' (*ITF News*, April 1998: 9). Outsourcing appears, therefore, to satisfy not only the company's cost concerns, but possibly also those relating to prerogative in the employment relationship as well as any interests in redefining employers' relationship with trade unions.

Cabin crew jobs have also been affected by outsourcing. At a time when safety experts are making it clear that the competence, co-ordination and training of staff must be given higher priority, it appears that airline work is being casualised and de-professionalised through an increasing use of casual and temporary jobs, and outsourcing. When, in December 1996, Air Gabon sacked its entire 94-strong cabin crew workforce to replace them with casual workers in order to cut costs,[4] an ITF representative commented: 'This is an industry in which neither professional nor safety standards can be sacrificed for short-term cost savings' (*ITF News*, January 1997).

Cabin crews have also been replaced in some European airlines (Britannia, Monarch, Air 2000) by 'flag of convenience' crews with lower training standards and more 'flexible' working arrangements, potentially providing a lever that may allow airlines to ratchet down the pay and conditions of home-based cabin crews. Hiring staff from abroad may reduce airlines' costs, but it has been argued, amongst other effects, that this practice can also create safety risks. It has been claimed that hiring cabin crews who do not have a good command of the English language has already led to serious safety problems – such as misinterpreting pilot instructions in emergency situations (Association of Flight Attendants 1997). Equally worryingly, cabin crew in seven developing countries do not even implement cabin crew training standards, yet can still operate UK flights under a 'flag of convenience' (*ITF News*, February 1998).

As many deregulated airlines have shifted the emphasis from service to cost, the value of the cabin crew appears to have decreased. In a contemptuous move against long-serving highly qualified cabin crew employees, BA carried out a cut-price crewing operation for new transatlantic flights. The airline went on to recruit cabin crews at pay and conditions far below existing BA cabin crew levels, leading to claims from BA staff that the company was virtually recruiting people off the streets and turning them into operational crews within days. Furious employees argued that BA's recruitment campaign made a mockery of the extensive training and experience of existing crews (*Herald*, 3 January 1995).

The professionalism of ground crews has also come under attack by some airlines' profit myopia. Aviation trade unions report that their members are under pressure to overlook safety misdemeanours to enable airlines to get their aircraft back into the air as quickly as possible (ITF 1998a). According to the ITF, this can be attributed in part to the emergence of anti-union, inexperienced, low-cost carriers, which makes it easier for managers to put pressure on employees to violate safety practices. At airlines where there are unions, the unions increasingly find themselves having to defend members who have been penalised by their companies for sticking to safety rules. The Fijian court that found Air Fiji to be 'overly concerned with economic considerations at the expense of safety concerns' was responding to the case of a ground-crew member who had been suspended for raising concerns about overloading on a flight (ITF 1998b).

In December 1997 the Australian civil aviation authority wrote to all licensed engineers saying: 'Sufficient anecdotal evidence exists to show that some organisations pressure licensed engineers to certify work that they would not have ordinarily have signed for, or to breach regulations in some way which affect air safety. Many maintenance engineers fear losing their jobs if they do not comply with such "requests" from their managers' (ITF 1998b). Subsequently, aviation trade unions joined the campaign for 'whistle-blower' protection, to ensure that responsible workers are not penalised for safety conscientiousness. In the UK, calls for 'whistle-blower' protection was addressed by The Public Interest Disclosure Act 1998, which provides protection for all employees, third-party contractors and the self-employed who disclose information about malpractice in their workplace. Unsurprisingly, all of these pressures and changes in the international airline industry have taken their toll on employment relations.

Employment relations

While traditional labour–management relationships produced an airline industry that has historically provided its employees with an unparalleled combination of benefits and working conditions, the markedly changed operating environment has effectively brought down the axe on such favourable situations. Restructuring in the industry has led to an assault on airline employees' pay, terms and conditions, leading to tumultuous industrial relations in a number of countries. Many of the trade unions' grievances can be linked to the cost-efficiency strategies employed by airlines, including outsourcing and swingeing cuts to labour costs.

After years of making substantial wage concessions to help airlines survive the recession of the 1980s and early 1990s, embittered employees began pointing to airlines' profits and demanding payback. Long-running disputes over pay and working conditions are visible around the world, and the call for solidarity amongst aviation workers is being acted upon. Just one month after the launch of the Star Alliance, the ITF announced the formation of the Star Solidarity Alliance. Cabin crew, pilots and ground staff in Canada, Australia, Britain, Spain, France, Germany, Hungary, Poland and Hong Kong put their names behind a united union agreement binding every union to share openly all information and details of collective agreements. However, in recent disputes over pay and working conditions, solidarity support has been obstructed by legislation in various countries. For example, the planned support from KLM pilots for their colleagues at Northwest in August 1998 was ruled by a Dutch judge as illegal. The expressions of solidarity mark a turnaround from the historic 'go-it-alone' strategy of airline trade unions (Moody 1987), as well as co-operation and solidarity across boundaries becoming something more than an idealistic notion. Without doubt, the backdrop of rapidly changing and threatening conditions since deregulation, in both the USA and Europe, has made airline unions more relevant to one another, and these trends are likely to extend into the future, placing a premium on unified efforts.

Many of these developments in airline employment relations are crystallised in the strike by British Airways' cabin crews in July 1997. While the crux of the dispute was over pay cuts and extended working hours, the related OHS implications of longer working hours on cabin crew health and safety were of considerable concern to union officials. BA's handling of the cabin crew dispute demonstrated the airline's readiness to abandon any people-centred HRM (and OHS) policy when it comes down to profit and survival. In addition, this example demonstrates the political basis of health and safety where assumptions pertaining to 'consensus' are quickly swept away. The BA cabin crew dispute also highlights the internationalisation potential of labour disputes as waves of support for the BA crew spread across the globe, offering some promise for future international solidarity against the range of OHS risks in aircraft cabins. Let us review some of the airline's machinations.

The BA cabin crew dispute

The origins of the dispute are located in a new deal for cabin crew pay and working conditions. The deal was part of BA's 'Business Efficiency Plan', which sought to cut costs by £1 billion per year. BA stated that £42 million of this would come from reductions to cabin crew costs (*Herald*, 10 July 1997). While basic pay would be increased, overtime and other allowances would be cut, duty times extended, turnaround times reduced, and new employees would start at lower rates of pay. Cabin crew estimated that they would lose thousands of pounds a year and would be working up to twenty-eight extra days a year for no extra pay. Adding insult to injury, the new-entrant cabin crew pay was slashed from £11,000 to £8,000 per year – a 20 per cent reduction – making BA cabin crew the cheapest in Europe (*ITF News*, 8 July 1997).

Despite the fact that the majority of BA cabin crews are represented by the Transport and General Workers' Union/British Airlines Stewards and Stewardesses Association (TGWU/BASSA) (around 9,000 of BA's 12,000 cabin crew), the company in the first instance approached and reached agreement on these new conditions with a much smaller union – Cabin Crew 89. This faction broke away from the TGWU in 1989 with, it has been proposed, the active encouragement of BA management. It is alleged that Robert Ayling helped set up Cabin Crew 89 when he was BA's Human Resources Director 'with the express intention of encouraging the kind of split in the BA workforce that led to the formation of the Union of Democratic Mineworkers which defied the 1984 miners strike' (*Independent*, 26 July 1997). Following the agreement of Cabin Crew 89, the company then attempted to impose this deal on all cabin crew staff. The TGWU refused to accept this imposition and balloted its members on industrial action. At the time of the dispute, David Cockcroft, general secretary of the ITF, commented that:

> This dispute is about basic rights. It is about a company saying to its workers 'we are going to increase your work hours and lower your pay, and take it or leave it'. When employees objected, the company responded with threats and intimidation of a kind which has never been experienced before in the airline industry.
>
> (ITF Media Release, 8 July 1997)

BASSA members voted 4,795 to 1,770 in favour of action in a 75 per cent turnout, and a three-day strike was called for 9–12 July 1997. BA management had made it known that, in the event of a strike, they still intended to operate by using managers, Cabin Crew 89 members and by recruiting a replacement workforce (*Observer*, 6 July 1997). They responded to the vote by issuing threats to sue, suspend or dismiss anyone who took part in the strike. At the same time, BASSA's offices at Heathrow were cleared and locked up by the company and, the union claimed, members were telephoned at home by managers. In the event, not only did the strike receive solid support from BASSA members, but around 2,000 of them submitted medical certificates, exploiting BA management's formal position (but one that is contradicted by punitive absence-management policies) that in the event of the slightest ailment, cabin crew members should not fly for fear of worsening their condition and/or spreading the infection onboard the aircraft (*Guardian*, 26 July 1997). BA was unable to provide the vast majority of its services, not only during the three-day strike, but also for several weeks afterwards as the cabin crews sat out their medical certification period. The 'mass sickie'[5] turned out to be as effective a form of industrial action, as strike action, since it grounded a substantial proportion of BA international flights.

The bully tactics of BA extended beyond its own employees to larger playing fields. When the ITF began to co-ordinate international union support for the strike, BA threatened the ITF with legal action. The company claimed that the ITF had breached UK labour law by calling upon foreign trade unions to support their colleagues in BA. In addition, ground staff in Zimbabwe were arrested after placing banners reading 'BA union-busters go home' on the check-in desks and in the BA departure lounge, and BA managers in Rome airport had Italian union members removed who were distributing leaflets to passengers at BA check-in desks. International solidarity was widespread, however. A picket at JFK airport in New York reportedly kept four jumbos on the ground and delayed at least one Concorde flight, while BA check-in staff in Paris (Roissy) went on strike for 72 hours in solidarity with the TGWU/BASSA strike. Cabin crew organised by the SNPNC in TAT, BA's subsidiary in France, also held lightning one-day strikes that further disrupted BA's global networks. The SNPNC had previously held a 28-day strike against BA following cuts to their pay and conditions, making them the worst in the French aviation industry. In addition, TAT ground staff organised a 7-day strike action. Other solidarity action occurred in various countries including Australia and New Zealand. BA worked hard to minimise the effects of the strike action, through, for example, hiring non-union crews. This, however, was severely hampered by continued shows of international solidarity. Charter carriers in Holland and in the USA, including Tower Air, World Airways and American Transair, declined BA contracts because of threats by the pilots and flight attendants' unions (for example, the AFA) in the respective countries that they would not accept strike-breaking by the company. Clearly, international unions recognised the trend-setting power of BA and the potential for the imposition of the new working conditions to have a knock-on effect in the industry, which is likely to have led to the rapid internationalisation of the dispute. However, in the end, many of the new BA conditions have been put in place, with the unions winning few concessions. Overall, BA experienced severe disruption to international flights and reports of the costs of the strike were estimated at around £125 million, plus the millions of pounds that were wiped off the company share value during the dispute (*Guardian*, 5 August 1997). At the very least, the strike cost BA many times more than the £42 million the company's imposed conditions were aimed at saving.

Thus, it is clear that the working conditions of cabin crews have been under attack from a sustained employer offensive. In all of this, of course, the experiences of the cabin crew bear a marked similarity to those of many other groups of workers across the globe during the 1990s, where all too often what is essentially a 'brutalist' strategy is dressed in the language of Human Resource Management (Warhurst 1995; Blyton and Turnbull 1995). Our expectations for the management of OHS under this type of regime are understandably subdued. This pessimism finds some justification when we consider the quality of the cabin working environment and cabin crew work organisation. In both of these areas, cost-cutting strategies appear as all pervasive, thrusting probing tentacles into every facet of cabin crew work in order to identify further opportunities to minimise costs, creating in turn a range of OHS risks that go well beyond the dangers of long working hours (Boyd and Bain 1998). Following recent public inquiries, the nature and extent of these risks is beginning to emerge, exposing, in turn, the long-term intransigence of the international airline industry in acknowledging the presence, never mind the associated OHS outcomes, of a myriad of occupational risks.

Air travel and health

Across the globe some 7 million passenger flights took place in 2000. For the UK, between March 2000 and April 2001 about 1.1 million passenger flights carried some 104 million passengers. While there is a range of economic and technical (that is, safety) aspects in current European aviation legislation, there is currently no legislation related specifically to passenger health. However, this is set to change with the European Commission consultation document published in January 2000 on air passenger rights in the European Union. This includes a section on conditions in the aircraft cabin. Another catalyst for change is the UK public inquiry into air travel and health (May 2000), which made a total of forty-seven recommendations covering passenger and air crew health (House of Lords Select Committee 2000). During the UK inquiry, a representative from the Joint Aviation Authorities (JAA) agreed that a body should be established that focuses on health issues on aircraft. In his words:

> We do not know enough about the interaction of large numbers of people on board aircraft . . . the industry is talking about loads of 500 to 600 passengers . . . in the end aircraft are going to be built which will carry 800 to 850 passengers. We have no knowledge what will happen to a group of such people travelling 12 hours in the same aircraft.
> (Klaus Koplin, Joint Aviation Authorities, 2000)

In the UK, a new department responsible for 'aviation health' is to be set up by ministers to investigate and regulate the industry amidst concerns over a variety of health and safety concerns related to flying. Following the Select Committee's inquiry (2000), it is hoped that some of the recommendations will appear in the proposed White Paper on Air Travel and Health. The basis of these recommendations is found in an examination of the range of OHS risks present in the aircraft cabin.

The cabin environment

> The aircraft is the place of work for the crew, to whom the airline has a duty of care as an employer. Exclusions from health and safety legislation are a legal technicality. A reasonable and prudent employer would take note of them and apply them in aircraft whenever practically possible, despite the exclusion.
>
> (British Airways 2000: 103)

If you were to think about your last journey on an aircraft, which features would spring to mind? The cramped, overcrowded space? The stuffy air? The smelly toilets? Thankfully, as passengers, we only have to bear these unpleasant idiosyncrasies of aircraft cabins for a relatively short time. Imagine if this was your workplace. As BA rightly notes, exclusions from health and safety legislation liberate airlines from the rigidities imposed by the employee rights contained in standard health and safety legislation. However, the extent to which BA and other airlines behave 'responsibly' is put into question by a range of literature on the cabin working environment and cabin crew working practices.

Cabin air quality

Whilst most people's concerns about aircraft safety understandably focus upon the seemingly increasing possibility of a horrendous crash, other more insidious and widespread threats to the health and safety of airline crews and passengers have largely gone unrecognised. Despite reassurances from the airline industry that cabin air poses no health risks to passengers or crew, a rising number of reports document a catalogue of symptoms suffered by aircraft occupants (Vasak 1986; Driver et al. 1994; Mcfarland et al. 1994; Kenyon et al. 1996; Smith 1996). Cabin air quality has become a major concern to flight attendants and to (some) official bodies. In February 1996, an ICAO paper concluded that: 'Poor cabin air quality is a double safety risk, affecting both the performance of the crew and the health and safety of everyone on board' (ITF 1995).

While airlines will generally seek to maintain industry standards and comply with in-flight safety regulations, 'grey' areas such as air quality and circulation rates are susceptible to cost-cutting – first, in terms of reducing the fresh air provision rate, and, second, in failing to properly maintain the air-conditioning system. The air-conditioning systems, as part of the environmental control systems (ECSs), on most commercial aircraft are based on the same general design. Since air at 30,000 feet cannot be breathed by human beings, it has to be compressed to a pressure that is life-sustaining. Simply, compressed air is induced by, and 'bled' from, the engines and then passed through an ECS, where excess water is removed and the temperature is regulated. The process of continually extracting fresh air and delivering it into the aircraft cabin uses up fuel and thus increases running costs. Recirculated air provides a less expensive option, where a proportion of the cabin air can be passed through filters and then re-injected back into the cabin.

Recirculating stale cabin air was not normal practice until the early 1980s, when airlines responded to increased oil prices. The growing acceptance of recirculated air has led to a gradual rise from an initial 20 per cent to a now-standard proportion of 50 per cent. This was one of two ways in which cost savings were achieved. The second method was the reduction in ventilation rates. McDonnell Douglas issued a report in 1980 proposing that the reduction of ventilation by 50 per cent (by running only two out of the three air conditioners on low power) would save up to 62,000 gallons of fuel per year per plane. Based on 1980 prices, this equates to savings of approximately US$1.20 per passenger per flight. Indeed, some models of new aircraft are fitted with fuel-efficient air-conditioning systems that automatically run on low power and switch off an air-conditioning pack when the passenger load is less than half. Aviation regulators have approved this move even though studies on Lufthansa aircraft have shown that that carbon dioxide levels double if airpacks run at 50 per cent capacity (Robertson 1989). In addition, Brundrett (2001) notes that CO_2 levels can be particularly high during take-off and landing, when power requirements reduce the amount of ventilation available.

These short-sighted cost-reduction exercises have been criticised by a number of commentators, who argue that the risks to passenger comfort and health outweigh the small savings that can be made through recirculating cabin air and reducing ventilation air flows. Currently, airlines are free to set their own rate of air recirculation, a licence condemned by Paul Halfpenny, ex-chair of the US National Academy of Sciences Committee on Airliner Cabin Air Quality, who argues that standards should be set and regulated instead of being left to individual airlines' discretion (ITF 1995). However, in terms of environmental concerns, the practice of recirculating cabin air finds some support. An Airbus Industries' representative recently commented that a return to 100 per cent fresh air would increase fuel burn by between 1 and 2 per cent (Thomas 2000). Given that aircraft are bigger polluters than motor cars, as recognised in the Kyoto agreement, increasing their emissions has clear environmental implications. Amidst these debates, the fact of the matter is that recirculated air would not pose a threat to passenger health if assurances could be made about the maintenance (and actual presence) of air filters in ECSs.

As studies on 'sick building syndrome' have shown, health problems related to recirculated air can be minimised by applying appropriate cleaning and maintenance regimes. In terms of maintaining aircraft ECSs, the correct maintenance and cleaning of filters is equally important for good air quality in the cabin. However, the minimal amount of time that aircraft spend on the ground, particularly during peak periods, places severe limits on the thoroughness of maintenance and cleaning. Clogged air filters in ECSs are reported to slow down ventilation flow rates by 20 per cent, which can lead to prolonged exposure to aerolised droplets (from coughing, sneezing, yawning), which in turn is linked to the transmission of tuberculosis (TB) and other infectious diseases on flights (including flu, conjunctivitis, viral pneumonia). Coughing, for example, produces around 100,000 particles, which can be dispersed over twenty rows.[6] In a scathing response to evidence provided by Farrol Kahn of the Aviation Health Institute, Dr Ian Perry, Consultant in Occupational Medicine, said that Kahn's claims that filters are faulty and clogged is 'a stupid statement' because filters are changed regularly, otherwise air would fail to circulate. This rebuke should be tempered by the fact that 'regularly' is according to the airline's convenience, and clogged filters slow down ventilation flows rather than stop them immediately.[7]

The link between ventilation rates and the transmission of infection and disease is strengthened by a body of international research. For example, US research found that on one flight at least six passengers were infected by a fellow passenger suffering from TB (Mcfarland *et al.* 1994). The risks extend beyond TB, as demonstrated by Donnelly (1997), who investigated the spread of gastroenteritis on two charter flights. The virus was contracted by 46 per cent and 38 per cent of passengers respectively, and the cause was not water or food. The investigation found that the virus was spread directly from person to person and through aerosolisation, aided by the air conditioning on the aircraft. The risk of contracting infectious disease is even more pronounced when the aircraft is on the ground for prolonged periods; for example, during delays the air-conditioning unit is switched off. Indeed, in their review of the transmission of TB on aircraft, the WHO recommended that in the case of ground delays of more than 30 minutes, provisions must be made to supply adequate ventilation in the cabin (WHO 1999). Wise to these risks, the USA has already implemented legislation that prevents passengers from being kept on a stationary aircraft for more than half an hour in order to protect against the airborne spread of disease (Dawood 2000). There is a strong case for similar legislation to be introduced unilaterally, not least to prevent the common practice of passengers being herded on to aircraft to free space at departure gates, even when air traffic delays are known in advance.

While some aircraft manufacturers (for example, Airbus Industries) are responding to these concerns by upgrading air filters in air-conditioning and ventilation systems on new aircraft (for instance the A340), older aircraft including the Airbus A300 series and Boeing 747/757 (two of the most commonly used aircraft type on long-haul flights) are still fitted with the lower standard filters. These can be upgraded to the higher standard, but this is subject to the prerogative of the airlines. Regardless of 'high' or 'low' standard air filters, our acceptance of recirculated air in the aircraft cabin should be tempered by the fact that even with the high standard filters at a 50 per cent recirculation level, some 476,000 particulates will escape into the cabin environment, compared to only 0.07 for fresh-air ventilation.[8]

In the UK, the CAA has no idea how many aircraft still have the low standard filters.[9] While demonstrating apathy and incompetence in its failure to record and monitor the types of filters used on UK-registered aircraft, the CAA confidently defends their use, arguing that aircraft filters are of the same standard and quality as those used in hospitals. However, such confidence in the ability of these filters to trap particulates is called into question by the findings of a survey carried out by the union representing hospital radiographers in Britain. As mentioned earlier, the survey found that a wide range of illnesses and ailments experienced by individuals in various areas of the hospital appeared to have been contracted through the circulation of bacteria through air-conditioning and ventilation systems (Society of Radiographers (SOR) 1991). Mind you, UK travellers should count themselves lucky to have any filters in the air-conditioning systems at all, since the CAA places no requirement on their fitment because they 'do not address a threat to air safety'.[10] This is convenient because, according to evidence provided during the Select Committee inquiry into Air Travel and Health, airlines do not routinely test the filters *in situ*, despite the manufacturer's recommendation that they should do so since the filters have a failure rate of 25–30 per cent for first-time tests.[11] So while filters may be installed, we have no idea how well they are functioning on flights (if at all).

Without getting too embroiled in the bickering about the efficacy of air filters, it is fair to say that much more research on cabin air quality is needed, and in particular on the analysis of 'real-time' air quality samples taken at different times during flights, on all models of aircraft with different passenger loads. At the time of writing, the American Society of Heating, Refrigeration and Air Conditioning Engineers (ASHRAE), which is responsible for setting air quality standards in the USA, was about to recommend a further reduction in fresh air, which will appear in a new standard for air quality within commercial aircraft (SPC161). Considering the range of risks that recirculated air and low ventilation flow rates produce, many passengers might happily pay a few extra dollars to minimise the risk of contracting various infections or experiencing a range of symptoms during or after the flight. For cabin crews, for whom this insalubrious environment constitutes the workplace, the impact of these environmental factors is likely to be far more significant. Fortunately for flight safety, pilots have a separate air-system cabin. A small mercy, considering the noxious cocktail that may circulate in the aircraft cabin.

The contiguous nature of working (and travelling) in the aircraft cabin means that the transmission of infectious diseases is potentially an everyday reality, particularly where cabin ventilation systems are under-performing. Given the obvious risks of disease transmission, it is astonishing that there is no international requirement for airlines to dissuade travellers from flying while they have an infectious disease (ranging from the more mundane viral infections to the more exotic such as TB). Furthermore, airlines are under no obligation to keep passenger information that would enable them to trace passengers following flights. The importance, and indeed, necessity of such procedures is demonstrated by one incident. Following an international flight, the airline Lufthansa contacted passengers to inform them that a TB sufferer travelling to the USA for treatment had been on the flight. It was reported that the airline did not normally keep this information and it was highly fortuitous that they were able to warn those travelling on the flight to visit their doctors for TB screening (Channel 4, *Dispatches*, 6 February 1998). This information led to two British women undergoing six months of preventative drug treatment, during which time they suffered a range of unpleasant side-effects, including nausea and headaches. While neither of the women contracted TB following the drug treatment, both were none the less furious that they had to undergo such treatment in the first place. Clearly, airlines have an obligation to refuse transit to people suffering from infectious diseases. This was another recommendation of the UK Select Committee, which, in keeping true to form, was dodged by the airlines. Consequently, the limited regard shown by airlines towards the transit of infectious passengers remains the status quo.[12] Remember that the next time you sit beside (or up to twenty rows away from) anyone who coughs, sneezes or yawns on the aircraft.

The UK government qualifies non-intervention in terms of international competitiveness. In its own words: 'The Government would be reluctant to impose obligations on UK carriers unilaterally because of the impact such action could have on their international competitiveness' (*The Government's Response to the Report of The House of Lords Select Committee on Science and Technology: Air Travel and Health*, 2001: 11). While the UK government tows the self-regulatory line and continues to bleat about the need for 'voluntary' responses from airlines, some positive, albeit minimalist, moves by some airlines are being taken. Although entirely vulnerable to the whims of the airline, BA has announced that it will retain passenger data for three months, while the ICAO has put forward a number of non-binding 'recommendations'.

While air quality and ventilation rates may be factors in the spread of infectious diseases onboard aircraft, cabin crew are also at risk of contracting disease by the very nature of their job. A number of occupational groups are exposed to the risk of infection, such as nurses, dentists, police, housekeepers and airline cabin crews. The risk of exposure to disease is increased as a consequence of performing routine duties when treating injury or illness during flights. In this capacity, cabin crew may be exposed to blood spills, vomit, urine, faeces, mucus and saliva. However, unlike the aforementioned occupational groups, many cabin crews around the world are not covered by a national training standard on incidental exposure to blood-borne pathogens, nor do they have the universal right to protective clothing, such as latex gloves, when carrying out routine duties.

And while the risk of contracting infectious diseases onboard aircraft is of considerable concern, there is a far broader range of contaminants in cabin air. These were highlighted during a legal case presented by US flight attendants against US tobacco companies in 1998. The tobacco companies' lawyers argued that other airborne pollutants such as ozone, cosmic radiation and non-tobacco contaminants had contributed to the various diseases (for example, lung cancer, heart disease) suffered by the litigious flight attendants. This is significant in that it is one of the first publicly stated acknowledgements of the damaging health effects of ozone and radiation to airline crews – areas that we will shortly explore. The outcome of the legal case was an agreement by the tobacco companies to fund a £185 million research foundation to investigate illnesses related to passive smoking onboard aircraft. The flight attendants retained the right to pursue damage claims of more than 3 billion dollars, and the airlines agreed not to block the rights of up to 60,000 cabin crews to file individual suits. Despite cigarette smoke being recorded as the most common complaint by passengers and crew members concerning overall air quality on flights (Oldaker and Conrad 1987), and a 1995 ICAO non-binding recommendation for a worldwide ban on smoking, many airlines did not ban smoking on flights until well into the 1990s. Consequently, some may find themselves drowning in a sea of litigation from vexatious employees – not least those suffering from the possible side effects of exposure to ozone and cosmic radiation.

Ozone

As argued by the US tobacco companies' lawyers, an additional hazard associated with aircraft cabins is ozone. Ozone is created from the effect of ultraviolet radiation on molecules of oxygen and is formed in the upper atmosphere, especially above 35,000 feet (the cruising altitude of many aircraft). There is clinical and laboratory evidence that exposure to ozone can cause pulmonary symptoms, such as coughs (Melton 1982). Other symptoms include headaches and loss of concentration. For those exposed to ozone, pulmonary symptoms occur at much lower concentrations with physical activity, hence flight attendants are much more likely to suffer from the effects of ozone than pilots. Moreover, the presence of airborne bacteria and viruses on aircraft makes the sometimes-elevated levels of ozone in the cabin of even greater concern, since the respiratory system is more susceptible to infection following exposure to ozone. Casting doubt upon assurances that ozone levels in aircraft are at a safe level is a major study, which concludes that the measurement techniques applied in most studies of cabin ozone concentrations were inadequate (Nagda *et al.* 2000).

Cabin ozone concentrations depend on ambient levels at given attitudes, but also vary with the type of aircraft, passenger load and characteristics of the air-conditioning system. In the USA, the Federal Aviation Administration (FAA) has advised that routes and cruise altitudes should be avoided where high levels of ozone are predicted to occur (FAA Advisory Circular AC 120-38 1980). Notice that the terminology is 'should' rather than 'must'. While such an imposition would cause a flight-scheduling nightmare, the fitment of catalytic converters in aircraft would maintain ozone concentrations in accordance with medically safe guidelines at all times. Some operators (for example, Icelandair) have voluntarily taken this step, and British Airways has ozone converters fitted on all long-range aircraft. However, that still leaves a lot of aircraft operating without any protective mechanisms.

Radiation

Air crews are exposed to cosmic radiation levels that are higher than the cosmic and terrestrial radiation levels normally encountered on the ground. Cosmic radiation is a mixture of various types of ionising radiation. A more familiar form of ionising radiation is that emitted by medical and dental X-ray examinations. The galactic cosmic radiation field at aircraft operating altitudes is complex, with a large energy range and the presence of all particle types. Since no single device, active or passive, can satisfactorily measure the whole range of particle types and energies (Bagshaw 1998), we might expect recommended maximum limits of exposure to err on the side of caution. Indeed, the general uncertainty about actual exposure levels for air crew is underlined by Goldhagen (2000), who points out that the theoretical basis upon which calculations of cosmic radiation exposure are made incorporate a plus or minus 20 to 50 per cent uncertainty measure.

The intensity of cosmic radiation increases with altitude because the atmosphere becomes thinner, making it less absorbent. The most occupationally exposed workers in the UK are air crew with 4.6 milliseiverts[13] per year (mSv/year) (Bagshaw, Irvine and Davies 1996, see also Schrewe 2000). This compares to an annual exposure of 3.6 mSv received by nuclear power plant workers, and should be levelled against the general consensus that the risk of developing cancer starts to increase at a dosage of 6 mSv/year (Building Research Establishment/Department for Transport, Local Government and the Regions (BRE/DTLR) 2001: 55). While there is no legislation addressing passenger exposure to radiation through air travel, a European Directive, effective from 13 May 2000, requires European airlines to assess exposure to radiation when organising air crews' work schedules and to inform crew of the potential health risk that their work may involve. In addition, special protection for pregnant crew members is applicable (European Directive 96/29) – procedures that were not previously in place for air crews other than those on Concorde aircraft.

There are a number of methods for measuring crews' radiation exposure, one of which is wearing individual dosimeters (which measure radiation exposure), (passive) dosimeters in aircraft cabins, or, more commonly, the use of computer software packages that calculate the average radiation exposure on given routes. In one study, passive dosimeters recorded a dosage that was 68 per cent higher than the estimated doses calculated by the software package CARI. This highlights the danger of the current reliance on software packages to estimate radiation exposure, as well as their unreliability in making assessments about risks to flight crews. An alternative and more reliable method of measurement has been developed whereby chromosome damage is analysed in the blood cells of air crew (Scheid *et al.* 1993). A number of studies using this method have found statistically significant increases in damaged chromosomes in air crew, with cabin crew having relatively more damage than pilots (see, for example, Lebuser, Krasher and Nubohm 1995). However, its invasive nature might make this a less attractive option to airlines and crews.

The need for some form of reliable measurement is underlined by a range of studies that suggest chromosomal damage (see for example, Romano *et al.* 1997) and cancers are more prevalent in air crews than many other occupational groups. In a cohort study of Air Canada pilots, Band *et al.* (1996) reported an increased risk of brain and prostate cancers, malignant skin melanoma and acute myeloid leukaemia. Similarly, a Finnish study of air crews found a significant excess of breast cancer in female flight attendants (1.9-fold) and bone cancer in air crew (15-fold) when compared with the national average. It concluded that 'ionising radiation during flights may add to the cancer risk of all flight personnel' (Pukkala, Auvinen and Wahlberg 1995). Lynge (1996) and Mawson (1998) also report an increased risk of breast cancer for female flight attendants. Some of these findings were recently corroborated by one study that confirms rates of skin cancer were ten times higher than expected with a 25-fold greater risk for pilots (Rafnsson, Hrafnkelsson and Tulinius 2000), while another large study demonstrated an increased incidence of myeloid leukaemia in pilots (Gunderstrup and Storm 1999).

While an abundance of studies point to increased cancer risks in air crews, a study by Irvine and Davies (1999) claim that there is no evidence that any incidence of cancers in pilots is linked to cosmic radiation. Their study is based on over 6,000 British Airways pilots who were employed for at least a year between 1950 and 1992. The authors make no reference to the effect of length of service on the incidence of the range of cancer studies, casting doubt upon the validity of their assertion. A similar conclusion is reached by Zeeb *et al.* (2002) from a study of over 6000 cockpit crew in two German airlines (even though an increased risk of brain cancer was identified). Rafnsson *et al.* (2001) underline the difficulty in proving a causal link between cosmic radiation and cancer incidence in flight attendants because of the range of confounding factors, which of course underlines the need for further research, particularly when crews of the future will fly at higher altitudes and will, consequently, receive even higher dosages of radiation (Goldhagen 2000).

Additional risks are also apparent for pregnant air crew – both pre- and post-conception. In a study of biological effects of cosmic radiation, Goodhead (1999) found that cosmic radiation could damage DNA and the developing foetus. In addition, it was proposed that the damage to DNA could cause genetic mutations in human egg cells and sperm cells. One US report on the levels of radiation on thirty-two flights concluded that for about one-third of the flights studied, the radiation exposure of a pregnant crew member would exceed the recommended safe limits for pregnant women (DOT/FAA/AM-92/2). More recently, Waters, Bloom and Grajewski (2000) reported similar findings in their extensive study of US flight attendants.

This presents an alarming situation for female flight attendants who may not be aware of a pregnancy, perhaps because of irregular periods, and, as a result, are unwittingly exposing their unborn child to occupational radiation. Moreover, should their child suffer any abnormalities or disease at birth (or in later life), parents would find it extremely difficult (if not impossible) to prove employer liability. While it is normal international practice to 'ground' flight crew when pregnancy is declared, clear and adequate information on the risks involved in continuing to fly has not always been made available to crews. Under these circumstances, some crew might delay announcing their pregnancy in order to avoid the loss of lucrative foreign trips, or the desk job they will have to endure for the duration of the pregnancy.

As well as the risks of cosmic radiation to pregnant crew members, higher than average miscarriage rates are reported for flight crews. One British general practitioner has argued that flight attendants suffer more miscarriages than other working women (Goodwin 1996). These concerns are underwritten by Vaughan, Daling and Starzyk (1984). In examining data on the pregnancy outcomes in various occupational groups in Washington State, they note 'a significant increase in the reporting of prior spontaneous foetal loss by women usually employed as flight attendants in comparison to a sample of all other women having live births'.

This finding was corroborated by Daniel, Vaughan and Millies (1990), who found that cabin crew reported miscarriage nearly twice as often as other women, though the risk was not so marked when flight attendants were compared with other *employed* women. However, 'a clinically significant pregnancy risk among flight attendants' was noted. More recently, a Finnish study reported a slightly increased risk of miscarriage among female cabin crew members (Aspholm *et al.* 1999).

In addition to radiation exposure risks, the pregnant crew member and her unborn child are at risk to other environmental hazards, such as electromagnetic hazards, noise, low-frequency vibration and decreased air pressure (and thus reduced oxygen concentration).

According to a report from the United States Army Aeromedical Research Laboratory, electromagnetic hazard in the cabin can harm the developing foetus. The paper concluded that 'the issues and concerns of pregnancy and flying are complex ... but the potential for foetal injury and death is real' (Mason 1994).

Just as flight attendants (when out of UK air space) are excluded from the HASAWA (1974), they are also excluded from the EU Noise at Work Regulations (1994). While the health effects of low-frequency vibration and noise from the engines in airborne aircraft have not been extensively researched to date, the literature shows that occupational noise is linked to deafness in infants, while low-frequency vibration is implicated in miscarriage and birth defects. For example, Scandinavian and Canadian studies (Härikainen-Sörri 1988; Lalande 1986) link a threefold increase in infant hearing loss to the exposure to occupational noise (as low as 90 dB) during pregnancy, while Flournoy (1990) reports that occupational exposure to low-frequency vibration increases the risk of miscarriages and birth defects.

While EC Directive 96/29 is welcomed, the increased information and protection it provides has come too late for thousands of cockpit and cabin crews who have been subject to the effects of radiation without being provided with adequate information, which would have enabled them to assess the risks. Moreover, the irresponsible attitude of airline companies and aviation regulators is demonstrated by their apathy in responding to the long-term concerns of aviation trade unions and their members about this issue. Since it is not (yet) possible to shield aircraft from cosmic radiation exposure, other controls are required, such as co-ordinating flights with solar activity – a move that would cause complete chaos for flight scheduling. However, as the BRE/DTLR (2001: 57) study notes, any serious risk-management strategy is likely to have a high cost, and consequently very strong epidemiological evidence is needed in order to persuade airlines and aviation bodies to take the necessary action. Given the range of factors involved in cancer such as lifestyle issues, as well as the high costs associated with major research projects, the prospects for generating this type of evidence are less than optimistic.

Frequent flyers

Barish (1999) argues that the regulations covering radiation exposure offered to air crews should be extended to regular business flyers whose schedules include a substantial amount of flight time at high altitudes. Data from frequent flyer reward programmes found that in 1996 approximately

435,000 people in the USA travelled distances and routes that, according to medical estimates, would take them over the recommended maximum legal exposure limits for members of the public (Barish 1999). Farrol Kahn, director of the Aviation Health Institute, argues that in order to avoid future litigation, companies that send employees on regular trips by air should be aware of the health risks and inform their staff accordingly (BBC Online Health 29 June 1999). One possible development might be the extension of the ICRP occupational limit of 20 mSv per year (20 times that of the recommended limit for the public) to business travellers, since travel forms part of their job.

The FAA recently proposed that frequent flyers on transatlantic flights may be exposed to the equivalent of 170 chest X-rays a year, putting them at increased risk of cancer (BBC Online Health 29 June 1999). Friedberg *et al.* (2000) calculate that a return trip across the Atlantic is equivalent to between three and four chest X-rays. Based on this, the costs of foreign travel and working in a 'glamorous occupation' could be considerably higher than we anticipated. How many people do you know that are aware of these radiation dosages and the range of health risks in aircraft cabins? By some accounts, the international airline industry has known for a long time yet has failed to adequately inform both its employees and its customers. Their reticence and the fact that improvements have only been secured by legislative intervention serve to dispel any faith that we might have had in airlines to act in a 'responsible' manner. Can we really expect anything better for the management of other risks that are emerging? One of the most worrying of these, as noted by participants in the UK Institute for Environment and Health consultation exercise (2001: 48), is the potential contamination of aircraft cabin air with lubrication oils (containing organophosphates). Before discussing this further, we consider the related risk presented by cabin altitude levels and the practice of disinsection (spraying pesticides) in aircraft cabins.

Pesticide/organophosphate hazards

In 1984, the WHO recognised that disinsection in the presence of passengers and crew had, in some cases, caused allergic reactions and asthmatic attacks. Despite this, disinsection is commonly carried out on aircraft, both when passengers and crews are onboard and when the aircraft is empty. On flights to a number of countries, including Australia, Trinidad, Tobago, Grenada, India and Uruguay, crew and passengers are directly sprayed with aerosol pesticides. Aircraft occupants may also be exposed to pesticides, without their knowledge or consent, as a consequence of empty aircraft being sprayed prior to passenger boarding. These pesticides will remain in the air-conditioning system, on the fabric of seats and on carpets and other surfaces. The practice continues despite a statement from the US Environmental Protection Agency's Director of Pesticide Registration during a

1994 US Congressional Subcommittee, where it was acknowledged that spraying occupied passenger cabins with aerosol insecticides could create medical problems for people with allergies, chemical sensitivities, asthma and other respiratory problems. The evidence of the damaging health effects continues to build. A 1999 report lists a catalogue of complaints from crew and passengers, including headaches, nausea, fatigue, memory loss, a reduction in cognitive skills and immune system depression (Riley 1999), while medical studies suggest a link between organic solvents and multiple leukaemia and breast cancers (for example, Dorgan *et al.* 1999). The WHO has conducted its own research in relation to disinsection in readiness for making proposals for inclusion in the revised International Health Regulations.

Other noxious fumes in aircraft may include organo-phosphate (OP) hazards. On 9 September 2001, an inquiry into passenger and air-crew exposure to toxic chemicals in UK aircraft was announced (*Observer*, 9 September 2001). The inquiry will examine the contamination levels in aircraft cabins from toxic organophosphate chemicals following reports of pilot incapacitation during flights and claims that *at least* 30,000 passengers a year were exposed to an OP level of acute toxicity (Balouet 2000). In a review of the literature on the health effects of exposure to OPs in a variety of workplaces (covering 158 reports published from 1965 to 1993), the following diseases were observed: skin-contact dermatitis, oil acne and photosensitive allergic dermatitis; testis benign and malignant tumours; respiratory system nasal symptoms, rhinitis, nasal mucosal tumour, laryngeal cancer, bronchitis, lipoid pneumonia, lung fibrosis, lung cancer, bronchial asthma and others, including possible carcinogenecity and a high incidence of chromosomal change (Karube *et al.* 1995; see also Blair *et al.* 1998).

The main risk appears to emanate from oil leaks from the engines into the air-conditioning systems.[14] The result of these incidents is that aircraft cabin occupants are exposed to irritants and neuro-toxins in combusted products of hydraulic fluid, or an aerosolised mist of hydraulic fluid. The vaporised toxic chemicals are then dispersed around the cabin by the air-conditioning system, where they are absorbed through the skin, mouth and nose. There is some evidence that the type of oil used (Mobil 2) causes the chamber seals to deteriorate before inspection intervals, which means that fluid can leak into the air-conditioning systems – a main reason why the military stopped using Mobil 2 in the 1980s, although the commercial airline industry has not followed suit. Engine-seal manufacturers estimate that, in normal operations, engine-seal failures will occur in around one out of every 22,000 flights, thus almost guaranteeing aircraft occupants' exposure to OP toxins at some point in time. If we apply the figure of 7 million flights in 2000, 318 events of OP exposure in aircraft will have taken place across the world in one year. Consequently, many thousands of people may have experienced an added extra to their in-flight service.

In January 2000, the FAA's formal acknowledgement of the potential for cabin air contamination was published in the Proposed Airworthiness Directive. Its publication was 'prompted by reports of smoke and odour in the passenger cabin and cockpit due to hydraulic fluid leaking into the auxiliary power unit (APU) and subsequently into the air conditioning systems' (65 FR 2555). Indeed, the US Association of Flight Attendants (AFA) identified 8,268 reports that mentioned 'smell', 'fume', 'toxic fume' or 'toxic gas' between 1 January 1986 and 7 March 2000 from the FAA's Service Difficulty Reporting Systems (including all aeroplanes, commercial carriers, freight operators, general aviation and charter operators). This is an average of 580 reports every year.[15] In tracking the problem, the AFA has collected data from engineer logs regarding oil loss during flights, as recorded by the flight deck, with reports of 'misting' in the cabin and cabin crews' experience of various symptoms. These included feeling intoxicated, headaches, dizziness, giddiness, loss of motor co-ordination (shaky hands), twitching or myoclonic tremors, rashes, eczema and upper respiratory symptoms (including sinusitis and rhinitis). Some flight attendants have reported long-term mental impairment, which they connect to misting incidents. In one airline studied by the AFA, there were approximately 6.8 such incidents per month. The MD-80 and BAe146 aircraft appear to be the worst affected by oil leakages (Witkowski 1999).

As already mentioned, the leakage of fumes containing OPs into air-conditioning units has been linked to several alarming incidents involving pilots blacking out or becoming seriously disoriented during flights. BALPA, the UK pilots trade union, is in contact with the lead solicitors in a legal group action concerning sheep farmers who are suffering chronic neurological damage as a result of the obligatory use of OP sheep dips. It is estimated that some 3,000 legal cases are being pursued by pilots and cabin crews in Australia, the USA and Britain for compensation for long-term neurological damage as a result of occupational exposure to OPs.[16]

Hypoxia

Aircraft cabins are usually pressurised to an equivalent outside atmosphere of around 8,000 feet. However, this represents a significant fall in oxygen pressure of some 28 per cent (Goodwin 1996). This could present a risk for pregnant passengers and crew given that pregnancy will increase the mother's oxygen consumption by up to 14 per cent. Several studies suggest that the foetus can cope with this level of hypoxia at normal cabin altitudes. However, these studies were carried out on sedentary passengers – not working stewardesses. According to Thiebault (1997), following on from Cottrell *et al.* (1995), hypoxia presents a risk to all cabin occupants since even moderate altitudes of between 5,000 and 8,000 feet may produce hypoxic symptoms. In their extensive review of the literature, the BRE/DTLR (2001: 26) report notes:

At 8000ft, it is possible for some people to experience mild hypoxia, the symptoms of which include impaired mental performance, reduced exercise capability, fatigue. Some individuals suffer mild hyperventilation, headache, insomnia or digestive dysfunction. The effects are not great . . . although accident risk could increase.

Flynn and Thompson (1990) further propose that the mild levels of hypoxia that cabin altitudes produce may be enough to bring about subtle changes in mental status, which can lead to erratic behaviour and loss of impulse control – providing a further insight into the current phenomenon of 'air-rage'.

Deep vein thrombosis (DVT) and air travel

The dangers of forming a blood clot in the form of a deep vein thrombosis (DVT) during air travel have been associated with long-haul flights, cramped seating, reduced legroom, dehydration, poor air quality and inadequate ventilation (Burnard 1999; BRE/DTLR 2001). Indeed, 10 per cent of DVTs are associated with air travel.[17] Several studies are now underway to identify and define the risk factors for thrombosis formation during long-distance air travel (Kahn 1999,[18] Burnand 1999[19]). In a review of over eighty medical references of Economy Class Syndrome (a misleading term because DVTs are not confined to economy class passengers), there are at least ten studies that demonstrate a direct link between air travel and blood clots. Of these, one is especially worth noting. A US research programme reported that 50 per cent of patients suffering blood clotting had recently travelled in the air for four hours or more. The symptoms usually appeared within four days of flying and 35 per cent of people had no predisposition to the condition (Mercer and Brown 1998). Ribier *et al.* (1997) also found that the condition – usually associated with older people with predisposing medical histories – may also occur in young individuals with no medical history. More recently, Scurr *et al.*'s (2001) study of the frequency of DVT associated with long-haul air travel in people aged 50 or over, reported a high incidence of symptomless DVT (where the clot develops but does not cause symptoms until some time after the flight). This study features in an extensive and meticulous review of the literature on air travel and DVTs carried out by the Building Research Establishment. In the report, it is somewhat reluctantly acknowledged that: 'If these results were reproduced they would clearly establish lengthy air travel as a risk factor for thrombosis' (BRE/DTLR 2001: 16).

The BRE/DTLR (2001: 21) report also notes that there are case reports of cerebral vascular (Pfausler *et al.* 1996) and peripheral arterial thromboses (Ashkan *et al.* 1998; Teenan and Mackay 1992) related to air travel, which further strengthens the association between blood clots and air travel. Indeed, there is a corpus of evidence that links potentially lethal blood clots with the numbers of hours spent in cramped aircraft seats. Some of this evidence is unwittingly provided by the airlines themselves. In a document written for officials involved in the Department of Transport's aviation review, the Air Transport Users' Council said the risk was of 'great concern ... there appears to be sufficient evidence to suggest a relationship between long-haul flying and DVT' (*Observer*, 5 August 2001). During the House of Lords Select Committee inquiry, it was noted how airlines and the CAA were quick to dismiss the abundance of anecdotal evidence on DVT and other risks (such as the transmission of disease), even though they had little on which to base this disqualification on. As Lord Jenkins pointed out: 'It is not the same thing to say that we have seen no evidence that there is a risk as saying that the evidence shows that there is no risk' (paragraph 15).

Following the inquiry, airlines have been instructed by the DETR to issue health warnings with tickets for long-haul flights – a recommendation that dates back to 1995 when it was proposed by the ICAO but rejected by the airlines and aviation regulatory bodies. This hardly qualifies as decisive action in reducing the incidence of DVT, supporting instead a 'business-friendly approach' towards OHS in protecting the airlines from expensive reconfiguration exercises that would increase the space between seats but would consequently reduce the number of passengers (fares) on each flight. However, the catalyst for change may be the multimillion-pound lawsuits currently being lodged against BA and Virgin Atlantic by DVT victims and their families (the seat pitch is below the comfort level of 34 inches in both airlines)[20]. However, to their credit, and perhaps not as cynics might argue that the impetus for the move is to improve their position during the impending legal proceedings, BA has agreed to take part in a medical research project by putting forward 1,000 of its frequent flyers – the first time a British airline has agreed to assist directly with such research (*Observer*, 14 April 2002).

While the Select Committee was resolute in its recommendation that 'airlines and their associates reappraise their current practices in relation to not only the provision of information for passengers but also the design of the cabin and cabin service procedures' (paragraph 6.31), the Government/CAA response was somewhat feeble:

> Should the outcome of [proposed] research reveal an increased likelihood of DVT from a wide range of transport modes, or that the condition is so widespread that it requires action as a general health matter, then such action will be considered.
>
> (*The Government's Response to the Report of The House of Lords Select Committee on Science and Technology: Air Travel and Health*, 2001: 8)

Moreover, the Committee's recommendation that airlines and their associates 'reappraise ... the design of the cabin' (paragraph 6.31) was completely ignored, while the cheaper, easier option of providing leaflets with guidelines to all passengers was relayed in great detail. With not a hint of embarrassment, the airlines are suggesting that passengers accept the risk as an unavoidable part of air travel and should also take responsibility for reducing this risk by doing exercises in their seats and wearing support stockings! This is a wholly inadequate response since not only do the recommended support stockings represent an additional travel cost to passengers (never mind warranting a call to the fashion police), but passengers are unable to control or minimise many of the actual risks that can lead to DVT, such as low humidity and low cabin air pressure.

The Select Committee did not miss the observed apathy of airlines towards protecting their employees and customers from a wide range of potentially serious OHS risks. The report states:

> Our concern is not that health is secondary to safety but that it has been woefully neglected. We welcome the belated acceptance by the Department of the Environment, Transport and the Regions (DETR) that it has the lead within the United Kingdom and we recommend the Government to ensure that concern for passenger and crew health becomes a firm priority.
>
> (House of Lords Select Committee paragraph 8.9)

The Select Committee also recorded its surprise at 'the lack of attention – by regulators, airlines and aircrew trade unions – to the health of aircrew' (paragraph 3.48). One key recommendation is that the 'present rules, agreements and attitudes regarding the monitoring and recording of general health of aircrew, over and above their fitness to operate, should be reconsidered urgently' (paragraph 3.48). In responding to this recommendation, the government and the CAA referred only to monitoring pilot health, ignoring cabin crews altogether.[21] This is unsurprising, however, since the CAA has no regulatory input into cabin crew health screening. Until recently, there was no requirement for regular medical checks for cabin crews, while pilots are required to undergo aeromedical examinations once a year until the age of 40, and every 6 months after that (JAR-FLC3). However, in March 2000 the European Commission brokered an agreement between social partners in the aviation sector, entitling mobile staff (that is, cabin crew and pilots) to a free health assessment before their assignment to duties, and thereafter at regular intervals.

From this review of the extensive literature on the health risks in aircraft, the aircraft cabin can be likened to a chamber of horrors, where a range of insidious risks are present yet their existence is continually denied and perhaps even concealed by airlines. While DVT has finally hit the headlines and public attention and pressure has now forced UK airlines to acknowledge the risk, other risks such as organophosphates are likely to linger in some dark corner for a while longer. Disappointingly, the angry mob is too easily pacified. The simple issuing of guidelines appears to have removed some of the urgency in the debate over the risk of DVTs without actually addressing the range of causes – such as limited legroom because airlines follow a sardine-like seating plan. While the newly issued warnings of DVTs from airlines are surely welcomed, one wonders about the adequacy (and ethics) of this minimalist move. For example, if we had been warned in advance of the increased risk of being trapped inside a burning aircraft because the narrow space between seats at over-wing exits severely restricts egress flows (Channel 4, 4x4, 23 July 2001), would airlines be similarly liberated from any responsibility for the deaths of trapped passengers?

Prospects

The UK Select Committee's recommendations for increased regulation, monitoring and research are imbued with a sense of urgency and, also disbelief that, to date, so little is known and even less has been done to protect air crew and passenger health. The confusion over who should lead improvements to passenger and crew health was voiced by Lord Jenkins during the inquiry:

> What has become apparent to us is that this [cabin crew health] is nobody's baby . . . that it [is] no-one's responsibility . . . the CAA clearly did not feel themselves in any way responsible for this . . . the JAA . . . are not an initiating body at all. To whom should we address our recommendations? Who is in charge, or is it nobody?
>
> (paragraph 509)

In response, Mr Chris Mullin, Parliamentary Under-Secretary of State, DETR, said that 'the buck stops with the DETR' (paragraph 509). The subsequent reports carried out by the Institute for Environment and Health and the Building Research Establishment (BRE) on behalf of the government were published in January 2001 and September 2001 respectively. The BRE report has been referred to throughout this chapter and it provides a comprehensive and thorough literature review of the identified risk areas associated with air travel. However, some of the commentary is conspicuously biased in terms of understating the extent of the risks or leaving out crucial information, such as the Pukkala et al. (1995) study on the prevalence of cancers amongst airline cabin crew. The report completely omits Pukkala et al.'s finding that female flight attendants were shown to have a significantly higher risk of developing breast cancer, and focuses instead on a secondary conclusion of the study – that the high prevalence of skin cancers amongst pilots was due to sun exposure (BRE/DTLR 2001: 56).

While the report identifies a range of high priority research areas, namely DVT, cabin air quality and jet lag, its recommendations arguably reflect the cost concerns of the airline industry. No mention, for example, is made of increasing the seat pitch from the CAA minimum standard of 26 inches. However, more encouragingly, the report makes some crucially important suggestions. In relation to cabin air quality, it is proposed that airlines carry out 'real-time' air quality measurements using portable kits to investigate the key cabin air quality parameters, such as concentrations of common pollutants and organophosphates, as well as measuring the blood-oxygen saturation of crew and passengers, ventilation rates and air movement. In addition, a survey of air-filter condition and maintenance is suggested, although this is listed as a 'low priority' (BRE/DTLR 2001: 2).

Overall, the report calls for more research into DVTs, cabin air quality, transmission of infection, cosmic radiation and jet lag rather than making firm recommendations based on the vast literature already available. While the authors of the report reviewed more than 2,800 research papers on these areas, the call for further research was made on the basis of a range of criticisms about the various methodologies employed in the reviewed studies. Despite the fact that all of these research papers appeared in peer-reviewed medical or other scientific journals, their cumulative findings were simply not good enough to persuade the authors to make the solid and decisive recommendations that could lead to the removal of some of the identified risks to crew and passenger health and safety. We will have to wait and see if any of their proposals transcribe into regulations. In addition, it remains to be seen whether this marks the end or beginning of scrutiny of OHS in the airline industry. The same tentative steps towards change are apparent in other countries. The US and European aviation regulatory bodies (FAA and JAA) plan to develop 'terms of reference' (TOR) that address all aspects of the cabin environment – air pressure, temperature, ventilation rate, humidity, carbon-monoxide levels, carbon dioxide levels, ozone level, particulates, biological aerosols, toxic gases and fumes, ionising and non-ionising radiation and the transmission of disease. This is an important first step in recognising that these factors should be considered in the equation of aviation safety. However, any subsequent recommendations and eventual regulations are just as likely to be influenced by the outcomes of proposed and ongoing research programmes, competitive concerns in the industry, pressure from trade unions and public expectations.

Thus, we can see the interaction and role of various macro- and micro-level factors affecting OHS in the international airline industry. From political frameworks to the goals and motivations of employers and trade unions/employee representatives, the range and extent of OHS risks facing international airline cabin crews appears to be unfolding, revealing a less than glamorous working environment. From the macro-environment of political frameworks, international standards and trade-union activity, we focus more closely on the micro-level of workplace OHS where we review a range of workplace factors and policies covering work organisation and the related impact of these on cabin crew health and safety.

Cabin crew work: myths and realities

> This is without doubt one of the dirtiest, most unhygienic environ-
> ments to work in. We are serving food and drink, clearing used meal
> trays, used hand towels, used sickness disposal bags, used nappies. We
> also handle dirty rubbish bags and we clean out dirty seat pockets, dress
> dirty toilets (renew toilet rolls and hand towels). All of these duties
> are done without adequate cleaning/washing facilities or protective
> clothing.
>
> (Boyd 1996: 35)

Of all workers in an airline, the flight attendant probably has the most
contact with passengers, acting as an ambassador for the airline. But
however much airlines rely on the courtesy of their cabin crew to win and
retain satisfied customers, it is the ability of the cabin crew to respond to
emergencies that constitutes their primary role onboard aircraft. Cabin crews
have been described as physical pieces of safety equipment on aircraft. At
30,000 feet, they act as the fire department, paramedics and may have to
deal with terrorists or hijackers. Cabin crew – or 'flight attendants' as they
are known in North America – constitute some 40 per cent of all airline
employees. In 1996, the cabin crew population numbered almost 200,000
within the European Union (Howard 1998), while, in 1999, around 25,000
cabin crew were employed in the UK alone (Transport and General
Workers' Union (TGWU) 1998). In popular culture, the job of the 'airline
stewardess' has been historically viewed as both glamorous and, in many
quarters, as something to aspire to. Such attitudes have also been commonly
associated with air travel itself. The remainder of this chapter may go some
way to dispel such myths.

Trends in cabin crew work organisation

The aircraft is a unique working environment. As prisoners in a metal tube
for an average of 10 hours a day (Boyd 1996), cabin crew cannot go for a walk
to escape work pressures, nor can they open a window for some fresh air.
Instead, flight attendants are confined in an often overcrowded, noisy
environment where they are under continual pressure to complete a given
range of service duties within a set timescale. The growing demands on cabin
crew is evidenced by the introduction of more and more passenger 'in-flight'
services in some airlines, increasingly crowded rosters and the reduction of
crew numbers on flights. While service demands may be fewer on short
flights and on low-cost carriers, cabin crews on these aircraft can operate
between three and seven consecutive flights in a single duty (TGWU 1998).
In these cases, the physical effort involved in catering duties (service carts
weigh up to 70 kg) is exchanged for longer assembly lines of the travelling
public that require welcoming smiles, safety demonstrations, someone to

complain to about trivialities and so on – all occurring in a 'Groundhog Day'-type scenario. An increasingly regular feature of the job is related to airlines' expectation that crews successfully diffuse and manage unruly or abusive passenger behaviour, which highlights the demands made on cabin crews' emotional labour on top of the demands made on physical labour.

These trends in work intensification in the occupation mirror those reported by Hochschild (1983: 122–3), where she describes how recession in the 1970s set US airlines on a course for 'cost-efficient' flying, while deregulation in the industry led to a sharp increase in the number of passengers. Along with the insurgence of 'discount people' and 'Greyhound passengers', deregulation marked the end of the golden age of air travel – an epoch when flight attendants made personal introductions to passengers, and long layovers between flights along with large flight crews were the norm. In one respondent's words:

> We had three flight attendants for seventy-five passengers . . . a social director who introduced each of the flight attendants personally . . . there was more of a personal touch. The plane had only one aisle, and we had berths for the passengers to sleep in. We used to tuck people into bed.
>
> (Hochschild 1983: 122)

The move to 'cost-efficient' flying changed this type of personalised service into one where

> We avoid eye-contact and focus on the aisles . . . people usually wait for eye contact before they make a request, and if you have two and a quarter hours to do a cocktail and meal service, and it takes five minutes to answer an extra request, those requests add up and you can't do the service in time.
>
> (Hochschild 1983: 122)

Service quality was not, however, the only casualty of 'cost-efficient' flying. The quality of working life has been similarly eroded by higher passenger numbers and service demands, along with shorter layovers between flights. The increasing intensification of cabin crew work is also evidenced by reductions in crew numbers. According to one union official for Pan American, 'if we had the same ratio [of flight attendants to passengers] now that we had ten years ago we would need twenty flight attendants onboard, but we get by with twelve or fourteen now' (Hochschild 1983: 124). These developments radiated from the USA across the international airline industry as deregulation and economic vicissitudes transformed airlines' operational strategies and working practices, leading in turn to a general worsening of working conditions and the pauperisation of a once-privileged occupation.

The primary research conducted by the author on three UK airlines provides an insight into these developments, where cabin crews report on the health and safety consequences of a range of policies covering work organisation and the cabin working environment. From a sample of 926 cabin crews in charter and schedule airlines (representative of age, sex, grade and length of service), a range of qualitative and quantitative data were generated from a self-report semi-structured postal questionnaire survey and semi-structured one-to-one interviews. Initial findings are reported in Boyd and Bain (1999), which detailed cabin crews' experiences and perceptions of a range of key workplace OHS issues. A full analysis of the data appears in Boyd (2001). Given the coverage of civil aviation regulations and the similarity of regulatory frameworks in the UK, the USA and Australia, a level of generalisation across countries is possible. We now go on to discuss some of the key findings.

Developments in working practices

Working time

Two important dimensions of working time are working hours and working patterns. Given the 24/7 nature of the airline industry, it is unsurprising to find that shift working is the norm. However, it is important to emphasise that shift work in the airline industry takes on a different form from that of other occupations, such as the police, fire or medical services, as well as production workers. For these groups, organised patterns of day, back and/or night shifts operate in rotation, usually from month to month or week to week. These arrangements allow the body and circadian rhythms time to adjust to changes in sleeping times. For cabin crews, mixed patterns of days and nights, with report times before 6 am and after 6 pm, are commonly rostered in a single week, causing severe disruption to sleeping and eating patterns (see, for example, Suvanto *et al.* 1990; Ono *et al.* 1991; Härma, Suvanto and Partinen 1994). The primary research reported that the vast majority of respondents routinely experienced these types of irregular shift patterns, with over three-quarters (76%) reporting that they 'often' worked mixed shift patterns entailing both very early and very late starts.[22] While much of the medical research on the effects of shift working identifies a range of ailments, the effects of erratic shift patterns on the performance and health of cabin crew have *not* been studied to the same degree.

Some indication of the health effects of irregular shift working is, however, provided by studies on jet lag, where symptoms include fatigue, reduced alertness, impairment of mental performance including memory loss, irritability, nausea and digestive problems. Indeed, the UK Select Committee inquiry into air travel and health acknowledged that 'it seems to be implicit that crews suffering from jet lag (or fatigue for any other reason) is a health and safety issue on the aircraft' (BRE/DTLR 2001: 83). Given the range of symptoms related to jet lag, its relevance to air safety is patently obvious. Indeed, Wagner (1996) concluded that EU flight-duty and rest times for pilots needed to be reviewed, based on the observed risks from subjective and objective fatigue measurements (see also Cauldwell 1997). While the literature review carried out on behalf of the DLTR argues that there is no definitive evidence of the magnitude of the accident risk (for example, accidents during flights or air crashes) as a result of jet lag, the report does note that 'there is a lack of understanding of important effects at the policy-relevant level: the health, safety and work performance of passengers and crew' (BRE/DLTR 2001: 84). A research agenda is therefore set.

Based on the symptoms reported in jet lag studies, it is unsurprising to find a large majority of cabin crew respondents reporting that they felt 'less healthy' as a result of mixed shift working, and almost three-quarters of respondents blamed the variety of health complaints they suffered on working patterns. Reported symptoms ranged from fatigue and eye/nose/throat irritations to sleeping and digestive problems. Statistical analysis found that a greater number of symptoms was experienced by respondents who regularly worked mixed shift patterns and who reported an increase in the volume and intensity of work. This suggests a linkage between illness, shift working and workloads.

The extension of working hours was a prominent trend noted by respondents. This related to fewer days off and shorter rest periods between flights. Subsequently, crews' monthly rostered duty hours had crept up towards the ceiling set by the CAA. In addition to this, respondents noted the considerable amount of working time that was not accounted for in duty-hour calculations. The most common length of duty was reported as between nine and twelve hours. However, this does not accurately reflect the actual length of their working day. As noted by Cappelli (1995), cabin crews' working hours incorporate more than flying time. Crew often report early for duty in order to prepare administration for the forthcoming flight and leave the airport long after the flight to carry out other (unpaid) administrative duties. As one respondent explains:

> The minimum rest period between duties is nowhere near sufficient. The 11-hour rest period is calculated from 10 minutes after landing till the next duty. This is grossly inaccurate – after landing, crew have not even left the aircraft in that time period. It is generally around 45 minutes to 1 hour after landing that crew leave the crew room (after counting money etc). This leaves 10 hours from leaving the crew room to the next duty. After the time taken to travel home, this leaves around 8.5 hours at home to rest, eat, prepare for next duty (socialise?!). It is impossible to be alert and refreshed for the next duty. Crew are lethargic and unable to operate competently after these rest periods which are often rostered.
>
> (Boyd 2001: 350)

Working patterns were also constantly mentioned as a source of disruption to family and social life. Airlines appeared to be taking the mantra of 'flexibility' to the extreme with last-minute roster changes being a common practice. The feelings of many respondents are captured in the following statement:

> This year is the worst in my twelve years as cabin crew for being under-staffed and having too many long flights in one roster. The shift patterns have been very irregular and it's not unusual to have your roster changed three times in one week, making your home life and social life suffer.
>
> (Boyd 2001: 349)

Working time is clearly extended by longer shifts and more working hours packed into a monthly roster, with 64 per cent of respondents reporting that working hours had increased in the past twelve months. In addition to this, workers' productivity during the working day can be maximised by ensuring that every pore of that day is filled with work demands. This was certainly true in the case-study airlines, where a large number of respondents reported that there was little or no free time during flights, or between sectors (turnaround periods) to rest, eat or drink. In their last full work roster, more than one in five respondents stated they had 'often' or 'sometimes' experienced turnaround times of less than 30 minutes, while turnaround times of 30–60 minutes were 'often' experienced by three-quarters of respondents. As one respondent explains: 'Turnarounds are quicker than ever – cabin crew don't even have time for a drink. It's just constant hassle and pressure' (Boyd 2001: 189).

In terms of general work performance and health and safety, there can be little argument that the interests of everyone aboard an aircraft are best served by having refreshed and alert crew. Indeed, the literature on the operational safety performance of other shift workers, or those working for extended periods (up to ten hours per day over several days), reports that individual performance deteriorated after an average of six hours of work. To mitigate these effects, rest breaks are legally provided for the majority of employees. However, despite the similarities in working patterns and the safety-sensitive nature of cabin crew work, the European Union's Flight Time Limitations Working Group did not consider it necessary to give cabin crews the right to rest/meal breaks. Policymakers seem to perceive cabin crews as superhuman beings who can function effectively in a safety-sensitive role, despite experiencing irregular shiftworking patterns and long working hours involving continuous and intensive work, during which time few, if any, rest breaks will be taken. However, as it turns out, cabin crew find these regimes both physically and mentally exhausting. According to respondents:

> I'm worried about being so tired ... I'm completely knackered just now. I've been on to crewing three times in the past month for rostering me illegal duties. I know my limits but they argue with me and tell me that it's 'legal' and the flight deck back them up. These arseholes should

try working the rosters they give us. That computer thinks I'm a relative a robosteward.

(Boyd 2001: 306)

We always experience delays then have too short turnarounds so crew don't have time to eat a crew meal or have a drink. It's like working in a sweatshop.

(Boyd 2001: 324)

Even if we do manage to have a break, we still have passengers coming into the galleys and pushing their call bells. Consequently, we never have a proper break – only 30 seconds to gulp down some food.

(Boyd 2001: 186)

In terms of the provision of work breaks for the cabin crew, there were huge differences in the three airlines' practices. The study found that 74 per cent of respondents in one airline enjoyed no rest breaks at all during an average duty but, overall, for the three airlines the proportion was 28 per cent. This distortion in the figures is explained by the fact that in one of the unionised airlines the union had negotiated the right to at least one fifteen-minute break. However, when asked if they had enough rest/meal breaks, 88 per cent of all respondents said 'no', although 'only' 60 per cent of crew on long-haul flights took this position. As one respondent explains: 'Some days you can go 6–7 hrs without having a chance to go to the toilet or have anything to eat' (Boyd 2001: 189).

In a ten-hour working day, would you be satisfied with fifteen minutes' break time? Moreover, would you have the energy to undertake the type of workloads detailed below?

Workloads

The cabin crew labour process has undergone significant increases in the intensity, volume and speed of work – trends mirrored by an international survey of aviation trade unions, where an increase in work intensity affecting all groups of aviation workers was reported (Blyton *et al.* 1998). The study identified a similar trend, with over three-quarters of respondents reporting an increase in the volume (77 per cent), intensity (81 per cent) and speed (79 per cent) of work in the last year. As one respondent complained:

The company wants to give more to passengers but won't put extra crew on to cover demands. Therefore, you end up so tired trying to complete all services – you hardly ever get a break. When you do have a break it is always interrupted by passengers. I've worked 11 weekends in a row – I requested a weekend off 4 months in advance and still didn't get it!

(Boyd 2001: 330)

Although crews working on all types of flights recorded similar trends, those on long-haul flights were generally at the lower end of the 'intensification' scale. The main reason for this was the extended time periods within which service requirements could be met. For short-haul crews, a manic work regime appeared to be the norm. According to one respondent:

> The short-haul operation matches organised chaos. We spend 8 out of 10 services running around like headless chickens. During turnaround we are treading on cleaners and caterers while getting the aircraft ready for the next sector. Our endeavour to satisfy the passengers means that we compromise safety. I frequently finish work looking and feeling like I have been dragged through a hedge backwards.
>
> (Boyd 2001: 186)

The intensity of work, particularly on short-haul flights, is partly explained by an increase in the number of services offered by some airlines. As discussed earlier, increasing pressures emanating from the effects of deregulation and intensifying competition have provided the impetus for airlines to adopt a variety of strategies aimed at increasing competitiveness through service quality. These measures include a greater focus on in-flight customer care and service. While the low-cost market is based on minimal in-flight services, many other airlines have chosen to increase the number of services offered to passengers during flights. One UK charter airline flying to Florida offers two meal services, a minimum of four drinks services, ice cream, scratch-cards, theme-park ticket services, and duty-free sales. This range of services underlines the increasing pressure on crews to generate revenue on flights. As respondents explain:

> We're under pressure to make as much cash as possible ... we've got targets that we need to achieve to avoid hassle. Often it's a choice between the crew getting ten minutes to eat something, or doing a drinks service which might bring in around £200. At the end of the day, if we had to do an emergency evacuation on landing, we'd need food and a chance to gather our wits during a 10- or 12-hour day [to perform effectively].
>
> (Boyd 2001: 198)

> It is a basic need to eat and drink during dedicated breaks on long flights. However this is not possible due to hectic service schedules. Cabin crew managers incur the wrath of management for not achieving service targets so 99 per cent of the time crew miss out on breaks.
>
> (Boyd 2001: 326)

> There are too many services on short flights leaving us just 'throwing out' the minimum to our passengers. There is a real sense of failure

when almost always we have to chop out some service to get finished in time. We are always fighting the clock and always rushing.

(Boyd 2001: 327)

The possible ramifications for safety are made clear by one respondent:

I think the Company places unreasonable emphasis on trying to achieve onboard sales targets at the expense of crew breaks, health and morale . . . I would question crews' reliability to successfully perform an emergency evacuation when they are so tired.

(Boyd 2001: 187)

Work intensification and safety implications are further highlighted by the reported reduction in the number of crew on flights. Indeed, almost three-quarters of respondents reported fewer crew members on flights compared to last year. Respondents reported flying 'a crew member down' on a regular basis, but said that they still had the same service schedule to complete, thus effectively intensifying their workload. As respondents explain:

Today my report time was 04.20. Every flight that went out this morning was a crew member down . . . I am sick and tired of working doubly hard to make up for a missing person . . . In this 8-hour and 45-minutes duty today we took a ten-minute break to eat our breakfast standing up.

(Boyd 2001: 196)

Jet Two is compromising safety standards at work because their only concern is to get the aircraft off the ground in time. I have operated too many flights with 5 or 6 crew where I have been the ONLY experienced crew member. If there had been an emergency, I wouldn't like to say how my crew would have coped due to their inexperience.

(Boyd 2001: 197)

These findings underline a variety of practices that extend the length and intensity of cabin crews' working day. As argued by Braverman (1974), following Marx, management (the capitalist) takes up every means of increasing the output of workers, such as enforcing the longest possible working day (through, for example, unpaid administrative duties before and after paid duty hours and minimum, if any, rest breaks during working hours). The fruits of these efforts include reports of increased work-related stress and pressure from 82 per cent of respondents.

The physical demands involved in the occupation combine with those made on emotional labour to produce a job profile that makes us feel worn out just reading about it. As explained in Chapter Four, the growing incidence of customer violence (verbal and physical abuse) appears to place a greater demand on interactive service workers' emotional labour, which in turn was linked to a higher incidence of complaints about work-related stress and the occurrence of stress-related illnesses. Similar findings are mirrored in the present study.

Emotional labour

> Over the years the airline industry has taught its cabin crews to be very subservient towards passengers. The passenger is 'always right'. The customer is fully aware of this and takes full advantage of the situation. They know they can say anything they want to cabin staff and get away with it, and they usually do. I have been employed as a cabin crew member for the past twenty-one years and I have had to suffer a range of indignant remarks and affront on a daily basis.
>
> (Boyd 2002: 158)

While Deery *et al.* (2000) report that high workloads and the speed or intensity of work increase the risk of experiencing emotional exhaustion, they also identify dealing with abusive customers as a key variable. The rising incidence of customer abuse in the airline industry is a matter of growing concern. Indeed, 70 per cent of respondents in the present survey believed that the number of abusive/disruptive passengers had increased over the past year. Two respondents explain the strains of the job:

> Passengers really annoy me. They just don't get that we are here to save their ass, not to kiss it. I feel that our service schedules are so jam-packed that we look totally unprofessional tearing up and down the cabin, flinging out meals, drinks, whatever. To passengers we are stupid girls, they have no idea of what we're really there for and certainly don't treat us with any respect. I'm sick of being treated like an idiot and a slave.
>
> (Boyd 2001: 190)

> Passengers are a health and safety risk. I was on a flight recently where a guy pushed one of the crew and told her to 'fuck off'. He was drunk and difficult to calm down. Nobody appreciates the shit we take ... It's usually too much alcohol or long delays. I dread getting on the aircraft following a delay ... you just know it's going to be hassle from start to finish.
>
> (Boyd 2001: 194)

One could argue that maintaining 'a cool head' simply shows professionalism and that managing customer diatribe is simply part of interactive service work. Questions pertaining to whether 'customers' have always been so unpleasant towards service employees, or whether a decline in service standards (for example, train and flight delays) and wider changes in society (such as unemployment and its effects) have led to the exponential rise in customer violence, are open to debate. Those at the receiving end may be more immediately concerned about the impact of these trends on their well-being. Without doubt, daily streams of verbal abuse will create some level of emotional strain. In addition, containing one's feelings in a workplace where there is 'no escape' only adds to the pressure. Being reduced to screaming silently in the confines of the aircraft toilet while passengers knock impatiently on the door (as described by one respondent) conjures up a quite unacceptable image of the emotional strain that cabin crew may experience on a daily basis. Here is how two respondents described the situation:

> We are taking more and more abuse from passengers as they are packed in tighter and tighter in the aircraft. People now suffer from 'seat rage' if someone reclines their seat – we have to deal with this.
>
> (Boyd 2001: 351)

> The number of abusive passengers has risen dramatically in the past year. Management does not realise the amount of abuse we take and are expected to take from passengers – is this taken for granted in other jobs?
>
> (Boyd 2001: 351)

While the airline companies had to different degrees provided some training on how to manage customer violence, many respondents described the training as 'inadequate', 'unrealistic' and 'dangerous'. Some respondents voiced concern over the fact that cabin crew members may be encouraged to enter situations under the false premise that they will be able to diffuse the situation.

The study, therefore, presents strong evidence of a higher volume and intensity of work (both physical and emotional labour), all of which generates concern for the safety and well-being of cabin crews and passengers – trends that may be mirrored across the international airline industry. While cost efficiency may indeed be an intended outcome of airlines' strategies, unintended outcomes may include declining service standards and the plummeting morale of workers. As one respondent explains:

> My job was totally different when I joined in 1982. We worked 3 flights a week and morale was much higher. Now crews work too long, unsociable hours with rosters being changed at short notice. Crews now work 4–5 days a week. We all gave a much better service to passengers when we were not tired or worked into the ground.
>
> (Boyd 2001: 329)

Other unintended outcomes may also involve the deterioration of the health and well-being of cabin crews.

Cabin crew health

> When I joined Jet Two, I was in perfect health. Since then I've had severe ear problems, kidney infections and an unidentified blister rash that covered my whole body . . . I now have constant back pains through lifting heavy atlas boxes/moving carts – even the catering men struggle with these. I have a constant cough and chest infection. I am covered in bruises from either carts or passengers, so much so that when I was wearing my swimsuit the other day, people thought I had been beaten up by my boyfriend!
>
> (Boyd 2001: 203)

The insalubrious cocktail of poor environmental conditions (for example, cabin air quality, low humidity), irregular shift work and high workloads might be expected to take a considerable toll on cabin crew health. It isn't a huge surprise, therefore, to find that over three-quarters of respondents believed that their health had deteriorated since they began flying, with a large majority reporting that they suffered up to thirteen different symptoms while flying (including headaches, eye/nose/throat irritations, ear problems, anxiety, tiredness) (Boyd 2001: 204). Based on the nature of these symptoms, it appears that just as aircraft share a number of design characteristics with so-called 'sick buildings' (for example, recirculated air), its occupants also appear to suffer from a similar range of symptoms. Other symptoms were also recorded, with 68 per cent of respondents reporting sleeping problems, 42 per cent suffering digestive problems and 57 per cent reporting that they suffer from anxiety. All of these correlated with length of service suggesting in turn that prolonged exposure to these environmental conditions and working patterns will produce a greater number of symptoms and illnesses. This range of symptoms is also similar to those reported in jet lag and shift work studies, and the propensity to suffer them was linked to the regularity of 'mixed' shift patterns. Some respondents noted their concerns about the impact of the job on their health:

> Whilst the company does pay lip service to our 'well-being', facts are kept under wraps – we hear nothing about the effects of radiation, breast cancer, deafness etc. Because of the low salary and health effects I won't stay in this job much longer as I'm too scared of radiation and damage to my immune system.
>
> (Boyd 2001: 217)

> Over the last year the increasingly long days have taken a toll on my health. I constantly feel tired and my first day off I walk around like a zombie. I keep getting sore throats and niggly little illnesses. Due to the stress of being tired small things become major things.
>
> (Boyd 2001: 217)

Over my 25 years of flying I have suffered the following: severe back injury, tendonitis in the thumb, kidney stone, sinus problem – all of which I feel could have been prevented by better management of crew health and safety.

(Boyd 2001: 218)

I have never been so unhealthy since I joined an airline. I was mortified to discover recently that I had extensive scarring on my lungs due to tuberculosis, which was contracted within the last ten years. I joined the airline in 1988.

(Boyd 2001: 218)

I lost a baby while at BA due to a high blood pressure related condition. My doctor suggested that if I want more children I should stop flying.

(Boyd 2001: 218)

Respondents attributed symptoms they suffered to a range of factors, including cabin air quality, work regimes and poor hygiene standards on the aircraft. Awareness of the risks present in the workplace was evident, with cabin air quality featuring as the greatest health and safety concern of respondents, followed by hygiene standards. Given the minimal time afforded to cleaning between flights and the fact that crews share toilets with hundreds of other people (not forgetting that these facilities are in continuous use over a prolonged period), respondents' dissatisfaction with onboard hygiene is understandable, with over half rating hygiene standards as 'poor', while similar judgements were made on the overall quality of the working environment, with only 8 per cent rating this as either 'good' or 'very good'. Further problem areas included training and absence management programmes.

Training

No real proactive health and safety programme appears to be in place for cabin crew. Most colleagues do not seem to have received any health and safety awareness training, not even during the induction course.

(Boyd 2001: 232)

Training is one of the key areas that can be subject to cost cutting in a fiercely competitive climate. As already mentioned, wide variations exist in the coverage of cabin crew training. In the UK, training requirements are regulated and monitored by the CAA. The majority of the training is focused on safety and emergency procedures (SEPs) such as inflating slides and rafts, operating doors, evacuation procedures, and dealing with fire and first-aid emergencies. At least two weeks of full-time training courses were operated by the case-study airlines. Annual refresher courses on SEPs (usually up to five days per year)[23] are also required. While SEP training is essential, the broader picture of cabin crew work necessitates the inclusion of training on safe manual handling, contact with body fluids and the skills required to defuse and manage a hostile or violent passenger. The survey findings show varying levels of disregard for all of these areas.

Remarkably, in an occupation where much of the work is concentrated on lifting baggage/bar boxes, pulling and pushing carts, attending to sick passengers (involving regular contact with blood, vomit, and so on) and contact with abusive passengers is not uncommon, many respondents have received no training in these areas. In addition, 60 per cent of respondents had not received training in dealing with body fluids, while 69 per cent had received only a brief outline on safe lifting/pulling/pushing techniques. Although restraint courses were given by two of the airlines at the time of the survey, as already mentioned, many of the recipients criticised the course as being unrealistic on the grounds that recommended techniques did not account for restrictive uniforms and crowded, cramped aircraft cabins. As one interviewee explains:

> We just had a restraint course to show us how to deal with trouble. It was a total waste of time. The idiots who took it had no idea about the cramped space in an aircraft aisle. A waste of my time. I got reported for complaining to the course organiser about the content of the course. Now that's on my record. It's pathetic.
>
> (Boyd 2001: 234)

It appears, therefore, that where training relating to OHS is provided it is inadequate, and that overall, little recognition is given to training in many of the areas relating to cabin crew comfort, health and well-being. These issues assume a 'hot-potato' status as they are passed to and from the CAA and HSE, with neither body taking on full responsibility. All of this adds to an expanding portfolio of OHS oversights and misdemeanours in the case-study airlines.

Absence management

One surprising finding from the survey was the relatively low absence rates (around 4 per cent). Given the high proportion of respondents stating that they regularly suffer from a range of symptoms, one might expect higher absence levels. One reason for this anomaly may be the range of absence-management policies utilised by the case-study airlines. The success of these is beautifully illustrated by the finding that 93 per cent of respondents regularly came to work even when ill.

The decision to come to work when ill is understandable when the convoluted and intrusive absence control mechanisms are examined. The procedure for dealing with absence begins with informing the crewing department of the intended absence. This department organises stand-by call outs and cabin crew work rosters. Cabin crews are required to inform crewing of their intended absence at least two hours before their rostered report time. On each day of absence they must telephone crewing to provide information about when they think they will be fit to return to work. Crewing officers are at liberty to call absent crew at home at any time of the day or night to confirm any of these details or to discuss future roster changes. Based on this requirement, absent crew must be contactable at all times. On the first day back at work, crews fill out an 'absence notification form', on which details such as the reason for absence must be completed. Return-to-work interviews took place for absences of more than seven days or when regular absences had occurred. In addition to tight control measures, the airlines utilised financial sanctions, namely the loss of flight pay, to deter absence. Flight pay is an hourly-rate premium and can amount to around 30 per cent of the basic salary every month. A further incentive was the avoidance of 'roster-wiping', where absent crews lose forthcoming lucrative foreign trips.

Less overt mechanisms were also in place, demonstrating the potential benefits of teamworking from a managerial perspective. Camaraderie is purposely fostered in crews in a number of ways, one of which is the design of duty rosters that aim to keep the same group working together for a block number of days. During this short period of time a sense of loyalty to the team tends to develop and respondents noted how this could act as a motivator for them to come into work even when ill. In addition, respondents explained that last-minute absences meant that a colleague would have to be 'called out' on stand-by duty – another factor that had an inhibiting effect on absence rates. These and other reasons for attending work when ill are shown in Table 5.1.

Attendance appeared to be a major concern for cabin crews and managers alike but from quite different positions of power. According to respondents:

> The Company expects us to be healthier than the average person and frowns upon sickness, when our job provides very unhealthy working conditions – very early starts, long duties, lack of quality breaks, working all night.
>
> (Boyd 2001: 341)

Table 5.1 Reasons for coming to work when ill

Reason	Percentage of respondents (more than one answer could be given)
Understaffing	34
Fear of reprimand	36
Worried about sick-leave record	65
Loss of income	38

Source: Boyd (2001).

My duty manager has told me that if I had any more sickness I would have to be examined by the Company doctor to check my fitness and health for this job. I was so scared that I came to work when I was ill with the flu.

(Boyd 2001: 341)

Effectively, crews may be penalised for taking time off work when almost every factor associated with the design and content of their work could be related to ill-health. Of course, the glaring transgression is the fact that the airlines are unfazed by the range of strategies that may impel sick employees to come to work even when ill, with what appears to be a limited regard of the longer-term effects on those individuals' health, or the more immediate effect on their performance at work (both in service and safety terms). Short-term profit imperatives may mean that tighter discipline is adopted in favour of the introduction of initiatives that address the possible root causes of illness (for example, workloads and working patterns). This logic operates in reverse to that contained in the already mentioned influential report that suggests that improvements to employees' working conditions and working environment can lead to a reduction in absence rates (EFILWC 1997). As such, the airline companies appear to follow a bizarre logic of imposing invidious absence management controls, while at the same time attempting to secure employee commitment to service and other business objectives.

The glamorous façade of the cabin crew occupation is surely now tarnished beyond recognition. A final and decisive blow is provided by the finding that almost three-quarters of respondents rated management's commitment to health and safety as 'poor'. Their judgement appears to be well-founded.

Conclusions

Airlines have been depicted as organisations that are dependent on their cabin crews in terms of the delivery of their product – the flight. Effectively, cabin crews act as ambassadors of the airline, and in delivering the company's product they engage in prolonged interaction with customers. Based on the evidence that customers' perceptions of service quality are affected by the nature of the interaction with front-line staff (Ashforth and Humphrey 1993; Peccei and Rosenthal 1997), it is understandable that some airlines describe their people management policies along the lines of treating employees as their 'most important asset' and caring about 'not just about how they work, but how they live'. Implicit in such a statement is an assertion that people are the key to improved organisational performance, and only by implementing people-centred, 'high-commitment' policies will organisations succeed in securing a competitive edge. HRM, as a management philosophy, is founded on such an assertion. Based on the rhetoric, it may be easy to assume that under HRM, 'good' health and practice is assured – simply because a safe and healthy working environment is one of the most basic requirements within sophisticated 'high-commitment' management. Not so, it would seem.

The apparent failure of some airlines to practise what they preach in policy statements communicates a disheartening message about the actual position of health and safety in management agendas. The research findings point to a range of issues that are not currently included in, or adequately covered by, existing regulations, not just in the UK, but on an international scale. These include training provisions, cabin air-quality standards, hygiene standards and working patterns/hours. In addition, the impact of intensive working patterns and workloads (as well as a poor quality working environment) on cabin crew health all appear to be grossly underestimated by airlines and regulatory bodies, based on the continuance (and even extension) of current policies, as well as the lack of regulatory intervention in these areas.

Given that the same number of passengers die *during* flights as they do in air accidents every year, it seems reasonable to expect equal importance being afforded to in-flight passenger health and flight safety (Kahn 2000: 45). Based on the literature, there appears to be an urgent need for adequate regulations covering 'comfort' issues, such as good air quality and hygiene. And, while trade unions at an international level have enjoyed some success in achieving various OHS improvements for cabin crews, it is clear that there is still a lot of work to be done. But what of the non-unionised airlines? There is some ground to speculate that similar problems will exist in both the non-unionised and the low-cost airlines, and may even be exacerbated in the latter group by even tighter cost margins.

In all of this, however, it is important to note that, at last, scrutiny is being directed to a range of very serious health and safety risks that airline crews have been subjected to for many years. Many of these OHS risks can be linked to a range of cost-cutting strategies imposed by airlines, while at the same time the airline industry has perpetuated an information blackout on some of these hazards (for example, cosmic radiation). Cynics might argue that the recent information released by airlines on the risks of deep vein thrombosis (DVT) is related to their concerns that recent media attention on DVTs might have a damaging effect on passenger numbers as well as the public exposure of this risk by the recent public inquiry. However, the extent to which the inquiry will secure substantive improvements remains to be seen.

If we return our focus to HRM, a hybrid form of HRM appears to be in place that demands from its staff loyalty and commitment to organisational goals without giving much in return – unless of course the risk of illness and injury is considered a desirable reward. Moreover, the observed approach to people management appears to be highly incompatible with what could be regarded as the primary foci of the industry, namely safety and service. Overall, the research presents a picture of cabin crew work that is contrary to the image created by the rhetoric of policy statements and marketing images. Cabin crew work, perceived by some as an 'exotic' form of service work (Warhurst and Thompson 1999), has instead been depicted as an intensive and demanding occupation that takes place in what has been described as a chamber of horrors. The combining forces of deregulation, the unequal distribution of power in the employment relationship and an insatiable mission for profit, have exerted a mutually destructive impact on the once preferential pay, benefits and working conditions enjoyed by cabin crews and other aviation employees. Consequently, the reality of cabin crew work, health and safety, as depicted by the research findings, clashes quite spectacularly with airlines' professed approach to people management. Overall, the range and extent of bad practice in the international airline industry is quite remarkable, although not as remarkable as the fact that airlines routinely get away with it – at least until a critical level of small events combine to trigger a tragedy, which, sadly to date, appears to be the only catalyst for reform and improvement in OHS across the industrial landscape.

6 HRM and OHS in the international call centre industry

> Every working day on average, five new contact centres are established in Europe, Middle East and Asia (EMEA).
>
> (Datamonitor 2001)

One of the most pervasive changes to sweep through the modern world has been the combined use of telephone and computer technologies in the organisation of work. The speed, ease and cheapness of transporting information across and between organisations, cities and countries has created a range of opportunities for both job creation and the quality of working life. The transformational potential of information and communication technologies (ICTs) fits well with the popular depiction of the 'information society'. For Toffler (1980) the information society represents a 'third wave' in economic development, following on from agrarian and industrial economies. Utopian visionaries such as Toffler (1980) and Stonier (1983) regarded the latest technological developments as propelling modern societies into a postindustrialist society, which held promises of sweeping away poverty and inequalities along with environmental and ecological problems. In addition, safety at work would be improved as machines took over the dangerous as well as monotonous tasks, allowing humans greater freedom in the creative and informative aspects of work (see, for example, Zuboff 1988). A more critical perspective is, however, adopted by authors such as Lyon (1988), who argues that in many cases the new technologies actually exacerbate, rather than ameliorate, long-standing class and wealth inequalities, leading to less, rather than more, opportunity, freedom and prosperity. A similar dystopian view of the impact of technology on work was proposed by Braverman (1974), who argued that employers' utilisation of technology is more likely to involve extensive job degradation and reduced worker autonomy. The literature suggests that during a period of prolific spread and growth of ICTs, many employers have failed to notice (or value) iron-clad opportunities for task enhancement, choosing instead to melt down and recast the potential of the technologies into Tayloristic images. This may create, in turn, the same people-management and OHS problems that have been historically recorded on the assembly lines (see, for example, Karasek 1979). While the propensity for serial Taylorism[1] across industries is damaging to optimistic accounts of the 'information society', it also raises concerns about the related impact on employee health and safety.

These concerns are realised in work-related illness and injury figures for habitual computer and display screen equipment (DSE)[2] users. In the UK alone, there are around 6 million habitual users of DSE (HSC 2000b). A 1999 study of over 3,500 habitual DSE users found that over half (55 per cent) of respondents had experienced symptoms of upper limb disorders, including repetitive strain injury (RSI) and shoulder pain (Institute for Occupational Medicine 1999). These symptoms were related to a range of factors in the organisation of work, such as the number of hours spent using keyboards, the length of time spent at the keyboard without a break and the pressures created by work. Another study reports that respondents who had undergone DSE training were significantly less likely to complain of RSI symptoms, stress and visual discomfort (HSE 1998a). These findings allude to an intimate linkage between people management policies (for example, training interventions, work organisation) and work-related illness.

The debates and concerns about the impact of technology, in particular ICTs, on the nature of work and employee health can be explored in the context of international call centres. Indeed, the mushrooming growth of call centre operations around the world is a testament to how employers, governments and society have embraced and utilised recent technological developments. Many of the trends we discussed in earlier chapters, such as sectoral shift, the feminisation of the workforce and complex employment patterns, are put into context within this setting. As such, call centres provide a gateway into a contemporary workplace as well as an opportunity to explore contemporary approaches to people management and OHS. Of particular interest is the facilitative role, and to some degree contradictory presence, of HRM policies and practices in an industry where a premium is placed on employee commitment, but where the nature of work is often characterised by tight control mechanisms and intensive work regimes.

As with airline cabin crews, the role of call centre agents is dictated by the immediacy of the production process and a dependency on employees' personal characteristics to deliver high-quality service. This is where the fruits of focused recruitment strategies and culture management programmes are harvested, and where various techniques aimed at maintaining their sweetness will be deployed. However, the spoiling effect of both physical and electronic managerial scrutiny, related in turn to intensive work regimes, presents a number of challenges to management (and HRM). Facilitated by various technologies, the tentacles of managerial control can delve and probe into every minute aspect of call centre agent work, offering instantaneous and meticulous detail on employees' activities. These relate to both the quantitative and qualitative aspects of call centre work, including, for example, the number of calls answered, call length, accuracy and adherence to call scripts, along with quality assessments of call handlers' performance on dimensions such as 'helpfulness' and 'enthusiasm'. Indeed, for some commentators this amounts to 'management control being rendered perfect' (Fernie and Metcalf 1998), conjuring up images of Bentham's panopticon[3] and *Star Trek*'s Borg Empire. Other commentators have vigorously argued against this presumption, pointing to expressions of employee resistance in some call centres and the range of improvements won so far by trade unions (Taylor and Bain 2001). This suggests instead that 'resistance is not futile'.

Some relief from the tension created between tight control measures and performance goals might be found in a close relative of HRM, namely high-commitment management (HCM). Wood and de Menezes (1998: 488) describe HCM as typically involving focused and highly selective recruitment practices – internal labour markets that reward commitment, teamworking, methods of direct communication, training opportunities and job security. This presents an interesting twist to the HRM story. Mainly observed in the manufacturing industry, HCM is normally associated with some devolvement of power that allows for increased employee discretion and autonomy. In the call centre industry, the level of scrutiny and control over workers remains intact, making the presence of HCM rather contradictory. It could of course be argued that there is nothing 'contradictory' about the operation of these strategies since HCM, like HRM, simply functions to contain employees within tightly controlled boundaries in the first place. All of these practices are likely to have some bearing on both the physical and emotional aspects of work and, in turn, employee health.

Following a similar structure to the previous chapter, we consider key trends in the call centre industry, where we identify a range of macro- and micro-level factors that contribute to the OHS landscape. At a macro level, we find national regulatory structures and government policies setting the scene and operational framework for the international call centre industry. At an organisational level, we explore linkages between employee health and a range of workplace factors such as work organisation, performance targets and the quality of the working environment. We begin with a review of key trends in the international call centre industry.

Trends in the international call centre industry

> There are at least thirty times as many computer telephonists as coal miners in Britain today; more people work in this [call centre] sector than in coal, steel and vehicle production put together.
>
> (Fernie 1998)

Call centre operations do not constitute an 'industry' in the commonly accepted sense, but it is considered sensible to refer to call centres as an industry for a number of reasons. As Belt, Richardson and Webster (2000:

382) explain, the call centre community often defines itself as an industry, based on both the range of legitimising activities such as annual national and international call centre conferences, and the high level of homogeneity within its labour force, applied technologies and organisational templates.

Call centre work has been estimated to be growing at 30 per cent per annum in the EU as a whole and as much as 50 per cent in the UK, which, along with Ireland, the Netherlands, Belgium and Denmark, is one of the most favoured locations for call centres in the EU (Huws, Jagger and O'Regan 1999: 39). This prolific growth is also evident in Australia, where there are more than 6,000 call centres and the related employment is expanding at a rate of 20 per cent per annum (Barker 1998). The number of call centres in the UK has been estimated at between 4,000 and 5,000 (IDS 2000). Estimates of the number of UK call centre workers vary widely, from between 225,000 and 420,000 (IDS 2000) to 1.1 million (IDS 1998). Similar variations in estimates are reported in the USA, ranging from around 4 million by 2002 (Datamonitor 1999) to 8.7 million (Roncoroni 2000).

The variation in estimates is partly explained by differences in the definitions of 'call centre'. Call centres take on a variety of forms, ranging from in-house establishments to independent operations that serve one or several organisations. While the evolution of call centres is marked by the use of voice-recognition technology and web-enabled multi-media facilities, the call centre remains fundamentally characterised by the integration of telephone and computer technologies (Taylor and Bain 2001: 40).

A broader definition of call centre work is provided under the heading of 'e-work'. 'E-work' is defined as 'any work that is carried out away from an establishment and management from that establishment using information technology and a telecommunications link for receipt or delivery of the work' (Huws and O'Regan 2001). In a recent study of the spread of 'e-work' and call centres across Europe,[4] the authors report that nearly half of all establishments in Europe (49 per cent) practised some form of e-work. Two distinct groupings of e-work were identified: e-work that was undertaken by the organisation's own employees (11.8 per cent) and outsourced e-work (43 per cent). Toffler's (1980) teleworker based in the 'electronic cottage' (that is, the home) was one of the least-practised forms of e-work (about 1.4 per cent). The range of outsourced e-work included call centres (26 per cent) and freelance work ('e-lancers') (11 per cent). The authors conclude that e-work is taking place on a significant scale in Europe and it is of sufficient importance to have a direct impact on employment practices and to affect indirectly the levels of employment in a number of regions.

A study of twenty-three different countries reports that call centre markets throughout Europe, the Middle East and Asia (EMEA) are rapidly growing towards maturity, particularly in the more developed Western and northern European markets, while in the less-developed markets in eastern Europe, expansion continues (Datamonitor 2001). Of particular interest is the apparent polarisation of the skills-base in the industry, with call centres in the more mature markets evolving into multi-media contact centres, while in the newer markets, such as those in eastern European countries, lower-grade operations predominate (Datamonitor 2001). This highlights key areas of importance for international policymakers in accommodating (and capitalising upon) the call centre boom. In addition, our attention is drawn to the prospect of lower-skilled call centre work moving from the more developed economies to cheaper and less legislatively demanding countries. This has particular ramifications for worker health since health hazards such as repetitive strain injury, back and neck problems, eye strain and a range of stress-related disorders are reported as being particularly acute in offshore data-entry facilities (Huws *et al.* 1999: 88). The extent to which OHS features in policymakers' agendas is, therefore, of interest.

For policymakers, tracking the movement of jobs and changing employment and skills patterns are issues of central importance, which all in turn feed into national economic policy and planning, aimed, for example, at safeguarding jobs and minimising social exclusion. Facilitated by fluid organisational structures and the transportability of information across cyber-highways, the global movement of call centre operations is a highly feasible option for transnational companies. Adding weight to concerns about a growing international division of labour (mirroring events in the low-skill manufacturing industries during the 1960s and 1970s) is a study of trends in call centre employment patterns across 206 countries. According to the authors:

> By the mid 1990s it had become apparent that the scale of cross-border relocation of information processing work was considerable, and that it extended beyond simple data entry to work involving voice telephony and e-mail communication, as well as higher-skilled work such as computer programming, accountancy and interactive work involving membership of virtual teams. It was also becoming clear that that this type of employment is extremely footloose, with companies engaged in a constant search for ever-cheaper sources of labour.
>
> (Huws *et al.* 1999: 56)

Wherever call centres locate, the implications for employment patterns, skills and OHS in the various host countries are significant. For developing countries, a number of opportunities and threats are evident. Huws *et al.* (1999) report that call centre growth areas include Latin America, in particular Mexico, where the flow of telecom and computer-based imports and investments correlates with the ongoing liberalisation and privatisation of the telecoms sector. While providing a gateway to the rest of Latin America, the major investment by a range of US companies is likely to be 'encouraged' by an enthusiastic Mexican government, a pubescent regulatory framework and lower labour costs. Indeed, by the year 2010, US exports to

Latin America are expected to exceed those to the EU and Japan put together (Huws *et al.* 1999: 49). The extent to which the subsequent job-creation exercise in Mexico represents expansion by US companies (that is, new jobs), as opposed to migrating operation and jobs, will be of interest. Fears over the longevity of newly created jobs in the call centre industry are equally salient in the more-developed countries. In parts of the UK where call centres have been the most important single source of employment in recent years (fuelled by a massive injection of some £138 million of EU funds), there are emerging concerns that the jobs created will be short-lived (Richardson, Belt and Marshall 2000).

The movement of jobs in the industry is perpetuated by ongoing 'beauty-contest' events organised by national governments and local authorities – a precedent set by the Irish Development Agency (Richardson and Belt 2001). Other European agencies soon followed suit, notably the Dutch Foreign Investment Agency and the British 'Locate in Scotland' initiative. Pan-European competition was intensified as other countries, including Belgium and France, joined in the fervour to attract call centres to their shores, bringing with them the promise of job-creation opportunities (Richardson and Belt 2001: 71). This is illustrated by Bain (2001: 11), who reports on the decision by Mercedes-Benz to locate its new pan-European call centre in Maastricht because of substantial local-authority (Limburg) financial inducements and the prospect of lower labour costs – about 60 to 70 per cent less than those in Germany. Economic-development agencies across the globe have subsequently tailored a range of policies to maximise their respective country's attractiveness as a call centre location. This includes low property costs, a range of generous fiscal grants and incentives with 'no-tie-in' options – a factor that, as we will see, is increasingly beginning to come back to haunt them. While one study reports that labour market skills take precedence over cost in decisions relating to call centre location (Huws and O'Regan 2001), it could be argued that the often basic skills needed in call centres makes this an easily satisfied criterion. The main point here is that job creation in itself should not automatically qualify for congratulatory back-patting from international leaders – the quality of the jobs, the level of skill development offered and the working terms and conditions of workers are essential elements in defining the range of benefits and opportunities provided by call centre expansion.

Broadly speaking, controversy over working terms and conditions appears highest in countries where call centre operations are not extensively collectively or legally regulated. On the other hand, controversy over call centre work is relatively low in countries where sectoral collective agreements exist (for example the Scandinavian countries) (*European Industrial Relations Review* 2000a: 14). The national regulatory structures of individual countries and the presence and strength of trade union organisations appear as decisive factors in the 'levelling up' or 'levelling down' of call centre working terms and conditions.

Ironically, international government policies may have created a Franken-stein's monster in the call centre industry, where a range of financial-assistance programmes and opportunities result in employers adopting a nomadic pattern of call centre relocation as they move from one oasis to another in the search for more favourable conditions. Given these circum-stances, it is pertinent to ask about the legacy of call centres. Will there be a mass population of multi-media enabled 'knowledge' workers equipped with transferable expertise and in-depth knowledge or will their profile be closer to low-skilled button-pushers? Will capital simply follow a perpetual migration pattern around countries governed by obsequious governments offering plentiful supplies of cheap labour while tolerating exploitative working practices? Given the statistics for injuries and illnesses suffered by regular DSE users, the legacy of call centres may be more than wrecked regional (and national) economies – extensive work-related illness and injury may be an unexpected codicil in the bequest.

Employee health and safety is a particularly controversial area in the international call centre industry. The relationship between OHS and regu-latory frameworks is highlighted by Huws *et al.* (1999: 88), who, as men-tioned already, report that health hazards such as repetitive strain injury, back and neck problems, eye strain and a range of stress-related disorders are particularly acute in offshore, non-unionised workplaces. Indeed, in assess-ing the relative merits of rival locations for data-entry work, employers are reported to be particularly interested in the local regulations governing working hours or other aspects of employment protection (Huws *et al.* 1999: 88). While it could be argued that the international ubiquity of a range of health hazards in data-entry and call centre work has more to do with how work is organised than its global location, the parameters for employment protection are, after all, set by the regulatory framework and, as such, the influence of regulation cannot be overlooked. In addition to the controversy over the movement of jobs in the industry is the issue of gender, where a range of divisions and inequalities in the international call centre industry appear to correspond to those present in other industrial sectors.

Gender divisions in the call centre industry

In the UK, a disproportionate number of female call centre workers are clus-tered into part-time, low-skill routine jobs such as mail-order selling, direc-tory enquiries and dealing with customer complaints, while men are more likely to be found in higher-skilled technical call centres such as computer help lines (Huws *et al.* 1999: 93). While this mirrors the gendered division of labour throughout the UK service sector, it also corresponds to global employment trends in call centres where women represent some 82 per cent of all global call centre workers (Call Centre News 1998). The gendered division of labour in call centres is also evident in Belt's (2000) three-country study (Ireland, the Netherlands and the UK) of female employment in thirteen call centres in three different industry sectors. The study reports that the variation in the proportion of female employment was strongly influenced by sectoral differences and the nature of the task. For example, in the financial services sector women represented between 70 and 91 per cent of employees, while in computer services firms women represented only 46 per cent of the workforce. In addition, men vastly outnumbered women in specialised technical roles making up the majority of employees on software 'help-desks'. Conversely, women were concentrated in work tasks requiring little or no technological knowledge, for example customer service and sales positions (Belt 2000: 13).

While the proportion of women involved in atypical employment may go some way towards explaining these patterns, a further explanatory variable may be that women are thought to be better suited to routine machine-paced work than men. Belt (2000: 16) reports how a number of managers in the case-study call centres believed that women tend to be more adept than men at 'handling' the most routine work, and were able to 'stick at the job' and 'keep themselves motivated' under pressure. In addition, she proposes that the low-skill end of call centre work has been defined by employers to involve the use of skills and qualities that women are seen to possess by virtue of being female (Belt 2000: 22). For example, it is suggested that women are more sociable and empathetic, and consequently, have better communication skills than their male counterparts. Such skills are particularly relevant for dealing with difficult or aggressive customers. As Macdonald and Sirianni (1996: 15) explain, 'women are expected to be more nurturing and empathetic than men and to tolerate more offensive behaviour from customers'.

For female call centre workers, it appears that their concentration into the lower end of the call centre industry spectrum is based on highly gendered perceptions of workers' skills and aptitudes. As a result, women may receive more than their fair share of boring, repetitive work, as well as potentially greater exposure to difficult or abusive customers. All of this is likely to be compounded by the wide use of electronic surveillance. Belt (2000) reports that the range of electronic monitoring and recording devices was less stringently used in those positions in which product knowledge and technical skills were of prime importance – an area where males tend to be concentrated. It is a sad turn of events for women whose acknowledged resilience, flexibility and self-motivation is rewarded by poor quality job design and possibly more intensive, pressurised and stressful working conditions. But just when we think women's predicament cannot get any worse, it does. If we consider the potential occupational health outcomes of boring, repetitive work, greater exposure to 'difficult' customers, performance targets and surveillance along with minimal control over work, we could quite safely place a bet on women experiencing a disproportionate risk of suffering from a range of work-related illnesses, including RSI and stress-related illness.

Overall, it appears that the stereotypical views of male and female attributes and characteristics in the workplace have remained intact in the 'information age'. As Stanworth (2000) notes, there is much evidence that old forms of gender segregation and inequality persist in the information economy, with female-dominated occupations still downgraded and regarded as low in status and vulnerable to automation, while the new high-status technical jobs are overwhelmingly occupied by men. However, men should not get too comfortable. Just as not all men are handsome, not all possess higher-order technical skills. In call centres and the interactive service sector at large, where the catch phrases 'the customer is always right' and 'excellence in customer service' prevail, men may find it necessary to develop and adopt typically typecast feminine qualities. It would be of interest to observe the extent to which this will challenge some managers' gendered perspectives of labour and whether any differences can be observed in the type and nature of control and commitment strategies that are applied to men and women doing the same job.

The range of influences that contribute to the OHS milieu in the call centre industry are therefore shown to mirror those discussed in the first four chapters of this book. At a macro-level, domestic and international governments and other policymakers set the context and framework for call centre operations, where often the quantitative objectives of job creation and inward investment have clouded qualitative issues such as skills enhancement and the quality of working life. While we can take comfort that not all call centres can be characterised as 'white-collar sweatshops', it is true that many still fit this typecast. The impact on workers' health has already been hinted at, pointing in turn to similar contradictions between some people management policies and OHS that we found in the airline industry. The next part of the chapter explores these in more detail. In doing so, we review the literature on a range of OHS risks in call centres, including those related to work organisation, control over work, emotional labour and the physical working environment.

Work organisation and OHS

Work organisation appears as a regular defendant in work-health trials. Within call centres, work is organised around a range of information and communication technologies (ICTs). As mentioned earlier, some commentators endorse the view that ICTs present a wide range of opportunities for job enhancement and creative expression for employees (for example, Zuboff 1988), while others (for example, Braverman 1974) argue that while these opportunities exist, employers' thirst for control over the labour process will mean that the technology will be utilised in a way that ultimately degrades and dehumanises work. Supporting the latter view are the monitoring capabilities of the new technologies to record and analyse electronic footprints throughout the working day. Giving weight to concerns over the work-intensification potential of the technologies is a range of critical literature on the manufacturing and service sectors. Within this, technology (that is, the way in which it is utilised) is often depicted as an express train to work intensification and emasculation (see, for example, Sewell and Wilkinson 1992; Oliver and Wilkinson 1992; Delbridge and Turnbull 1992; Danford 1998; Taylor and Bain 2001). However, these technology-facilitated demands represent only one dimension of work intensification. It is now generally recognised that workers' emotional labour is part of the labour process, where the conscious management of emotions becomes a critical component of work (for example, in the interactive service sector). As such, significant demands may be made on workers to display the 'company face' when, for example, dealing with difficult and/or abusive customers or, less dramatically, quelling feelings of boredom and disinterest when dealing with monotonous and routine customer enquiries. Call centre workers may, therefore, experience a double whammy of work intensification and corresponding OHS outcomes. Later in the chapter we explore the subsequent demands on emotional labour and their related OHS outcomes in call centres, but for now we focus on a more general understanding of the nature of work organisation in call centres.

Call centre work organisation

Taylor and Bain's (1999) characterisation of call centre work as an 'assembly line in the head' captures the populist white-collar factory depiction (and sometimes reality) of low-skill, repetitive and monotonous work in some call centres. In obtaining a better understanding of the way in which work is organised in call centres, the 'quantity/quality' dimensions applied by Taylor and Bain (2001) are particularly useful. The authors provide a number of indicators, as shown in Table 6.1, that relate to the priorities of employers.

The 'quantity' dimension

> At present staff [call centre agents] . . . feel they are nothing but battery hens.
>
> (Quoted in Deery *et al.* 2000: 25)

The quantity dimension of call centre work organisation is well illustrated by Taylor and Bain's (2001) case study of Telecorp. In this call centre, tight call-handling times were recorded for various services, including up to 30 seconds for directory enquiries, 48 seconds for operator services and up to 100 seconds on international lines. These services were characterised as simple and repetitive work that was highly monitored (Taylor and Bain 2001). As already mentioned, the range of available technologies provides employers with almost unprecedented scope in monitoring workers' activities to the smallest details. Once logged into their computers, automatic call-distribution systems divert calls to the agent, while other systems record the length of

Table 6.1 Ideal characteristics of quantity/quality

Quantity	Quality
Simple customer interaction	Complex customer interaction
Routinisation	Individualisation/customisation
Targets – hard	Targets – soft
Strict script adherence	Flexible or no scripts
Tight call-handling times	Relaxed call-handling times
Tight 'wrap-up' times	Customer satisfaction a priority
Statistics-driven	Statistics modified by quality criteria
Volume	Value

Source: Taylor and Bain (2001: 45).

time taken to answer each call, the duration of each call and the amount of time not logged into the system, for example during toilet breaks. Tight call-handling times and monitoring combine with performance targets to accelerate the pace of work. According to one respondent:

> We have targets for everything; for how long the calls take, how long you take to type up the call, anything you can measure statistically. If it moves they'll measure it including how long you take to go to the toilet. We now have errors targets which are extremely tough to meet and put a tremendous strain on you.
>
> (Taylor and Bain 2001: 52)

As with airline cabin crews, a range of other factors contributes to the pace and intensity of work, including the number and length of breaks. In terms of breaks during working hours, a TUC report states that in some cases call centre agents were given only 75 minutes of break time in a 12-hour shift, while others agents claimed to get no breaks at all (TUC 2001b: 5). As well as creating greater demands on physical and emotional labour, extended periods working with visual display units (VDUs) can cause eyesight problems and headaches. One particular problem is that blink rates have been shown to drop when using a VDU. Consequently, the eyes are less well lubricated, and this in turn can lead to tired, sore eyes and headaches.

In its second report of the call centre workers' campaign, the TUC recorded call centre workers' main complaints as few or no breaks at all during working hours and extreme monitoring. Of those callers who complained about the level of monitoring, 53 per cent referred particularly to the monitoring of the number and length of toilet breaks. Some had to ask permission to go to the toilet and others complained of being hauled up in front of bosses to explain why they were going to the toilet so often. Call handlers from two separate call centres said managers had brought in disposable nappies to work saying that those who went to the toilet too often would have to wear one (TUC 2001b: 4).

Employers' failure to provide adequate breaks during working hours is a direct contravention of the Working Time Regulations, which state that workers are entitled to a 20-minute break away from their workstation every 6 hours. But compliance to legislation does not appear to be a strong point for many call centre employers. The HSE (1998a) reports that out of the organisations surveyed in relation to the implementation of the Display Screen Equipment (DSE) Regulations, only half had got anywhere near full compliance. Despite being in place for a number of years, the Working Time Regulations and the DSE Regulations appear to have had minimal effect on curtailing employers' enthusiasm for, and ability to impose, intensive and unhealthy work regimes in call centres. According to one report employers perceive little or no benefit from the DSE regulations in terms reducing employee absence or increasing productivity (Institute for Employment Studies 1997). This might go some way to explain their demonstrated contempt for them.

The 'quality' dimension

At the other end of the quantity/quality continuum, Lankshear *et al.*'s (2001) study of one call centre in the travel industry noted the particular influences of 'quality' dimensions in work organisation. In this study, customer transactions were complex and often time-consuming, involving the compilation of information relating to holiday bookings (for example, availability, insurance and prices) using computer-based reservations and booking systems. A recurrent theme during the case-study research was the value of quality over quantity. In the words of one agent:

> I don't care within reason how long I spend on a call. If a customer has gone away happy and booked something, then if I have spent half an hour on a call then I don't worry about it. You can't in this job just get rid of people quickly. To me customer care is more important than how quick you are doing it.
>
> (Lankshear *et al.* 2001: 601)

The respondents' views were consistent with organisational objectives on providing a high-quality service, and customer satisfaction was a priority. This was in turn supported by 'soft' targets, relaxed handling times and the absence of scripts. Indicators of employee satisfaction were relatively low turnover rates (15 per cent) and reported unsolicited positive comments from the call centre agents. The respondents' attitudes towards electronic surveillance and monitoring were of particular interest. Supporting Taylor and Bain's (2001) arguments about the weaknesses of exaggerated claims regarding the 'panopticon' effect of the surveillance technology in call centres, the authors note that despite the possibility that taping was taking place, call centre agents were undeterred in breaking company policy rules during bookings. However, these expressions of resistance to formal rules operated well within managerial boundaries, as they could be qualified on the basis of providing customised and enhanced customer service.

Clearly, a great deal of variation in working practices exists in the international call centre industry. The 'quality-orientated' operations appear to embrace some of the 'high-commitment management' practices mentioned at the beginning of the chapter, while lower down the call centre spectrum, storm clouds gather over employee health and safety.

Work organisation and OHS

While the research evidence on OHS issues in call centres is fairly thin on the ground, there is growing evidence that DSE work (particularly within intensive work regimes) is related to a range of injuries and illnesses including musculoskeletal disorders (MSDs) and hearing and voice loss. A Brazilian call centre study reports a corresponding increase in call centre agents' workloads and health problems, with about 80 per cent of respondents experiencing some kind of discomfort or pain associated with MSDs (Sznelwar, Mascia and Zilbovicius 1999: 295). Similarly, Most (1999) reports that half of the 1,312 survey respondents suffered some level of pain while carrying out their job. The pain was located primarily in the neck (50 per cent), shoulder (49 per cent), lower back (34 per cent) and wrists (26 per cent). Sore eyes were also reported by 18 per cent of all respondents. Giving some indication about workers' expectations is the finding that 97 per cent of these respondents believed the organisation was a safe place to work.

In addition to MSDs, hearing problems and voice loss are key OHS risks in the call centre labour process. A 1999 Royal National Institute for Deaf People (RNID)/Trades Union Congress (TUC)-commissioned survey of noise-induced hearing loss reports that acute noises and prolonged high levels of sound through headsets experienced by call centre workers could cause dulled hearing and/or tinnitus. In the survey, some respondents reported experiencing these symptoms for short periods after finishing work, while others reported permanent impairment or discomfort (Labour Research Council 1999). The report highlighted the negligence of call centre employers in alerting employees to the possible risks of noise at work. In fact, 90 per cent of call centre workers had not received any information or training about the risks of noise. As John Monks, General Secretary of the TUC, noted:

> We may have lost much old industry, but that does not mean noise at work is no longer a problem. Call centres are as noisy for their workers as many old style factories. Thousands of call centre workers, many in their twenties, are already losing their hearing thanks to unsuitable headsets.
>
> (TUC 1999c: 1)

However small-scale the RNID/TUC survey, its findings have nonetheless been acknowledged by the Health and Safety Executive (HSE) who also support calls for further research in this area. The increasing number of out-of-court settlements involving call centre agents for work-related hearing impairments points to the increasing prevalence of an uncontrolled risk in the call centre labour process. Equally, dysphonia is a growing problem. Dysphonia is not just the inability to speak, it also includes pain, tension, croakiness, irritative cough, inability to modulate, poor or no vocal power and breathing difficulties. In the UK, the HSE has been alerted by ear, nose and throat (ENT) surgeons about the increasing number of cases of voice loss being presented by call centre employees. Consequently, a nationwide survey of ENT surgeons is planned (HSE/HELA 1999: 10).

The experience of MSDs, hearing and voice loss may all be associated with prolonged and intense work. The role of electronic monitoring both in supporting the work regimes and also on the negative health outcomes is highlighted by Most's (1999) study. In his study of communication workers at US West Communications, he reports that the use of electronic monitoring systems increased workloads and work pressures to the degree that they increased workers' experience of work-related illnesses and injuries. In addition to these findings, a UK study found that monitored employees report higher levels of stress than unmonitored employees (HELA/HSE 2001: 16). Interestingly, Most (1999) reports that monitoring was not thought to improve work performance or motivation, suggesting that management's reliance on such systems within call centres is misplaced or is grounded in a different agenda altogether.

In common with other industries is the malaise of under-reporting evidenced by the finding that a minority of respondents in both Most's (1999) and Sznelwar *et al.*'s (1999) studies reported their symptoms. In Most's study, one of the main reasons for doing so was 'fear of job loss', bringing to mind the linkage between job insecurity and work-related illness (Mayhew and Quinlan 1999). In some cases the outcome of 'struggling on' might be that some workers are eventually debilitated and no longer able to work or able to seek appropriate compensation from their employer. Few opportunities may exist to seek reprisals (for example, through employee tribunals), leaving the individual dependent on state support (assuming this is available). Meanwhile, the employer will probably be recruiting the next batch of call centre fodder.

Work organisation and emotional labour

The demands on emotional labour are well illustrated by Deery *et al.*'s (2000: 24) observations:

> Invariably [the call centre agents] were the subject of customer irritation when they were neither the cause of the problem nor in a position

to remedy the customer's complaint. Service operators frequently had to confront customers who were hostile about telephone accounts that were incorrect or higher than expected; telephones that were disconnected because of non-payment of bills; slow installation of telephone equipment and computer difficulties and systems failures that sometimes impeded the location of relevant data or information.

Confronting perplexed and anxious callers is more forcefully highlighted by Taylor and Bain's (1999) study of UK emergency services (999) telephone operators. Call handlers would invariably be dealing with distressed callers who were requesting fire, ambulance or police emergency services. Some insight into call centre agents' work experiences are provided by the findings of TUC research. The TUC's 'It's Your Call' hotline was available to call centre agents over a two-week period in February 2001, allowing them to report on work grievances. Complaints about abusive customers featured prominently. Of particular concern was an expectation that employees should 'grin and bear it'. As one agent reports:

> There is no policy for dealing with abusive calls other than that call handlers are not allowed to hang up on anyone – no matter what the circumstances.
>
> (TUC 2001b: 4)

Even if those circumstances involve sexual harassment? It would appear so. Callers to the same hotline describe their experiences:

> We have an adult channel and have to take calls from people wanting to subscribe ... There have been a few times when I've taken orders from men and I can tell that they are masturbating while they are on the phone to me.
>
> (TUC 2001b: 4)

> I've seen young women working on this line who are completely distraught because of the abusive calls they get. Just last week, one colleague who is only 19 was literally almost sick.
>
> (TUC 2001b: 7)

A German study of 106 call centre employees reports that the experience of sexually harassing calls was 'extremely stressful' to some call handlers. In the study, three out of four females reported sexually harassing calls, with characteristic patterns of harassment including groaning, sexual insults, silence and threats of sexual violence (Sczesny and Stahlberg 2000: 121). These types of responses are recorded in other studies where call handlers experiencing sexual harassment have felt angry, uneasy, nervous, disgusted and degraded, while some experience trembling and increased heart rates (Clark, Borders and Knudson 1986; Katz 1994; Sczesny and Stahlberg 1999; Sheffield 1989; Smith and Morra 1994).

Sczesny and Stahlberg (2000: 123) report that gender was an important variable in the experience of sexually harassing calls, in terms of exposure to calls of this nature and the likelihood that the experience would be considered stressful. Women were more often sexually harassed over the telephone than their male counterparts and were more likely to report the experience as stressful. In addition, the stereotypical harassment pattern – a man harassing a woman – was overwhelmingly manifest. With regard to consequences, these experiences and the anticipation of them affected both job satisfaction and job performance in a negative way, suggesting broader implications for organisational performance (Sczesny and Stahlberg 2000: 133). The authors conclude that a similarly high occurrence of sexually harassing telephone calls in specific work environments with intensive employee telephone use, such as call centres, is likely – a finding that contradicts Pease (1985), who argued that sexual harassment on the telephone is rarely found in the workplace environment.

Emotional labour and OHS

While the outcomes of high demands on physical labour may include MSDs, the outcomes of high demands on emotional labour are reported to include emotional exhaustion, also referred to as 'burnout'. Burnout was first investigated in the caring professions, where the management of emotion is a central part of the work. In these professions, relationships with patients, clients or children are very demanding and require a high amount of empathy and emotional involvement (Maslach 1982; Maslach and Lieter 1997; Schaufeli and Buunk 1996; Schaufeli and Enzmann 1998), while the costs of these demands included psychosomatic complaints, depression and other long-term stress effects (Zapf *et al.* 1999: 372). Zapf *et al.* (1999) analyse emotion work (that is, work that involves emotional labour) from an industrial psychology perspective, arguing that emotion work is a source of psychological job stress, but one that is not incorporated into traditional concepts of psychological job stress.

The physical aspects of work equally influence the experience of emotional exhaustion. Deery, Iverson and Walsh's (2000) study of five Australian call centres in the telecommunications industry reports a statistically significant (p<0.05) relationship between the experience of emotional exhaustion and repetitive tasks/high workloads. As the authors explain:

> There were a number of job-related aspects of the call centre work that were associated with higher levels of emotional exhaustion. When the service operator's job involved more repetitive tasks and when the operator believed that their workload was excessive and required them to

work 'very fast', there was a significantly higher probability of emotional exhaustion or burnout.

(Deery, Iverson and Walsh 2000: 25)

Interestingly, the authors did not detect any statistical differences between men and women in terms of burnout or emotional distress – a finding that is consistent with Wharton's (1993) findings. In addition, the earlier suggestion that women are 'better equipped' to manage repetitive and intensive work (both in terms of physical and emotional labour) does not appear to be supported by these findings.

Unsurprisingly, the experience of emotional exhaustion has a positive association with the frequency of employee absence (Saxton, Phillips and Blakeney 1991; Deery, Iverson and Walsh 2000). This association could be explained in part as a response to debilitating health effects, and by the role of absence in providing employees with a safety valve. Hackett (1989) describes absenteeism as a form of 'exit' involving an escape from unpleasant working conditions, while Brooke (1986) characterises absenteeism as a pain-reductive response to a frequently uncongenial work condition (quoted in Deery, Iverson and Walsh 2000: 13). According to one respondent in Callaghan and Thompson's (2001) study:

Call centres have the worst rate of sickness and non-attendance due to stress. You don't want people shouting at you all day. Your line is active at all times. That, I believe, is more strenuous than physical activity.

(Callaghan and Thompson 2001: 33)

In Deery *et al.*'s (2000) study, higher levels of emotional burnout (or emotional exhaustion) corresponded to a significantly higher incidence of one- and two-day absences. Interestingly, this pattern was highly prevalent amongst those workers with high ratings for 'positive affectivity' – relating to personality traits such as a positive outlook on life and work (see, for example, Agho, Mueller and Price 1992). Deery *et al.* (2000: 27) explain that although employees with high positive affectivity may be less prone to emotional burnout, they may hold higher expectations about the intrinsic rewards they seek from work, and the lack of fulfilment of these rewards may lead to greater absenteeism.

Given the working conditions within some call centres, one could argue that it is hardly surprising that high employee turnover rates are a trademark of the industry, particularly in the UK. One of the most comprehensive studies of UK call centres reports that staff turnover was considered problematic in four out of ten call centres (IDS 2000). The average reported staff turnover across all of the 250 call centres surveyed was 20 per cent for permanent staff. In four out of the ten call centres that gave a turnover rate for temporary staff, the figure was 50 per cent. Regional differences were apparent, with companies in Scotland most likely to say they had a problem with staff retention, while those in Wales were least likely to claim so (IDS

2000). However, the degree to which high turnover was considered undesirable is questionable. According to one call centre manager:

> To me attrition is very healthy in a call centre, very costly, but very healthy. Because of the stressful nature of the job and because you want these people to keep constant energy and enthusiasm, it does the organisation some good if you can pump in some fresh blood.
>
> (Callaghan and Thompson 2001: 33)

For the less bloodthirsty managers wishing to reduce turnover rather than depend on a constant supply of new donors, taking action to resolve some of the reasons for leaving would be a first step. The IDS (2000) study reports that the most commonly identified cause of staff turnover was the intensity of the work. Further reasons included competition for staff from other call centres. The study further reports on some of the measures taken in some call centres to address staff turnover. These included enhanced training and development, improved communication with employees and greater employee involvement in decision-making, along with a more participative management style. In addition, better physical working conditions and greater variety in work were proposed. Many of the proposed remedies are simply 'good practice' both in HRM and OHS terms. However, given the considerable delay in making improvements to call centre working practices, it may be the case that 'good practice' is only introduced after the (cost) benefits of 'bad practice' have been exhausted.

The working environment and OHS

Notwithstanding the detrimental impact of some work organisation policies on call centre workers' health, the physical working environment in call centres has also been identified as host to a range of OHS risks. This is a primary finding in two influential UK reports on call centres. Both the 1999 and 2001 HSE/HELA[5] studies identify air quality, temperature and relative humidity as key problems reported by call centre workers. The 1999 report notes:

> Such a high concentration of people and continuous occupation [in call centres] increases the risk of high levels of germs and volatile organic compounds (VOCs) ... both of these could lead to sickness absence. VOCs are also released by certain glues, paints and carpets, and, with such rapid growth, many call centres are in buildings that have been recently constructed or refurbished where these materials have been used.
>
> (HSE/HELA 1999: 3)

Temperature problems in particular call centres further highlight the relationship between OHS issues and building design. The location of air vents appeared to be an invisible factor in the layout of the office space. Vents were often blocked by screens, resulting in pockets of stale, hot or cold air, and subsequent discomfort to the office occupants. One example cited by the HSE/HELA report (1999: 4) describes how ground air vents in one building were designed to cool temperatures at both ground level and the level above. However, call handlers situated near these vents on the ground floor were too cold and so covered the vents with boxes and rubbish bins. This resulted in the mezzanine floor becoming unacceptably warm. The report also identified a range of sick building syndrome-related symptoms including sore eyes, sore throats, voice loss, headaches and migraines, highlighting the combined influences of environmental conditions and working practices on call centre workers' health.

Conflicts and contradictions

The 'quality' and 'quantity' emphases of the people management policies employed across the call centre industry create a number of contradictions. As Pfeffer (1997: 115) notes:

> One might wonder, in an era in which there is increasing emphasis on building employee commitment ... how there can be a concomitant increase in electronic and other surveillance activities – a practice that research indicates often produces effects opposite to those being sought.

Kinnie, Hutchinson and Purcell (2000) report on how HR practices play a crucial role in solving this conundrum. The authors quote Schnieder and Bowen (1993: 40), who assert that 'what employees experience at work gets transmitted to customers ... the key to managing the customer's experience of service quality is to manage employees' experiences within their own organisation ... HR management is crucial'. Kinnie *et al.* (2000) go on to discuss two case-study call centres based in areas with rapidly tightening labour markets and where call centre agents' jobs were characteristically tightly controlled, allowing little discretion over work. High customer-satisfaction levels through improved service quality were shared goals and their realisation was sought through the introduction of high-commitment management (HCM) practices. As mentioned earlier, Wood and de Menezes (1998: 488) describe HCM as typically involving focused and highly selective recruitment practices, internal labour markets that reward commitment, teamworking, methods of direct communication, training opportunities and job security. In the manufacturing industry, HCM is normally associated with an increase in employee discretion and autonomy, leading Kinnie *et al.* (2000) to comment on the contradictory nature of HCM within call centres. The authors describe how a range of HCM practices was introduced including teamworking and teambuilding strategies, and pay and recognition systems that combined successfully to engender a level of employee commitment to customer service quality. In their words:

Employees were encouraged to take responsibility for their own performance, and that of their teams, and make suggestions for improvements. The pay systems were changed to reward high performers and to encourage more short-term incentives, and teamworking was introduced. Teams were not required by the technical aspects of the job since it could be carried out independently, but instead they were used to manage individual performance more effectively, to seek to exercise normative control over attitudes and behaviour and to provide a social dimension both at work and occasionally outside work. In addition, there was careful attention to recruitment and selection, extensive training and development, including the use of personal coaches, the establishment of some type of career structure and attempts to improve working conditions.

(Kinnie, Hutchinson and Purcell 2000: 981)

As a result, labour turnover tumbled from between 27 per cent and 35 per cent during the previous three-year period to 8 per cent, while absenteeism halved (Kinnie, Hutchinson and Purcell 2000: 978). The suggestion is that employees' commitment to customer service quality was won through the use of financial bonuses, recognition in the form of 'spot prizes' (for example, chocolates, bottles of wine), improved training and a friendly working environment (for example, direct communication, themed fancy-dress days). Underneath these decorative embellishments, however, lies the same demanding, pressurised and low-discretion work. As one call centre manager commented:

It's a very structured environment – both physical and mental constraints. The fact that they have to perform all the time is quite stressful. The fact that they have customers shouting at them, members of staff being unhelpful is stressful . . . and of course the IT is stressful.

(Kinnie, Hutchinson and Purcell 2000: 979)

As Taylor and Bain (1999: 110) note, 'even in the most quality driven call centres it is difficult to escape the conclusion that the labour process is intrinsically demanding, repetitive and frequently stressful'. However, perhaps there is nothing 'contradictory' about the operation of HCM strategies within tightly-controlled work regimes since HCM, like HRM, acts to contain employees within managerially-defined boundaries of control, discretion and responsibility. Perhaps employee commitment in some call centres really comes down to how well the blinkers, as provided by HCM, fit. While Kinnie, Hutchinson and Purcell's (2000) case study aligns with Lankshear *et al.*'s (2001) depiction of a call centre in which 'quality' is emphasised over 'quantity', the next example exhibits a more cynical application of HRM policies and practices.

The sacrificial HR strategy

The term 'sacrificial HR strategy' fits well with the patterns observed in some call centres relating to job design and work organisation, the utilisation of technology, as well as trends such as high staff turnover and employee absence. Based on patterns identified in four large Australian call centres, Wallace, Eagleson and Waldersee (2000: 179) describe the 'sacrificial HR strategy' as an approach to people management that recognises the tensions between cost control and service quality but does nothing to solve these tensions. Instead, this approach relies on the strategic recruitment of service-orientated, motivated workers, whose skills and effort are swallowed up by the company. Burned-out waste products are easily expelled from the organisational body via high attrition rates. According to one manager: 'We don't want people to stay past 18 months. By that stage they are burnt out and are no good' (Wallace, Eagleson and Waldersee 2000: 178).

The 'sacrificial HR strategy' is one, therefore, that aligns with accounts of 'hard', cost-rational HRM, and quite audaciously, accepts and expecting employee burnout, stress and high turnover. In the words of the authors:

> There was a clear understanding of the savings that could be gained by turning over burned-out staff rather than investing in programs targeting morale, commitment and enthusiasm. By deliberately selecting individuals whose intrinsic motivation was service, high service levels were assured and the need for the organisation to provide this motivation ... was removed. By accepting burnout and high turnover, there is a reduced need for the organisation to manage the [outcomes of demands made on] emotional labour.
>
> (Wallace, Eagleson and Waldersee 2000: 179)

Interestingly, the median length of service for call centre agents was 15 months, which would appear to fit with call centre employers' goals of 'flushing out' or refreshing their workforce at regular intervals, and would also support the suggestion that after this amount of time, call centre agents are emotionally burned-out and no longer achieve performance targets.

Key contingencies for such an approach to function 'effectively' are likely to include a plentiful labour market and the design of tasks such that product and process knowledge is embedded into the system (that is, the technology) rather than the individual. This results in minimal training costs and enhanced managerial control over work processes, where virtually every activity performed by the employee can be monitored and recorded. These conditions are amply provided for in many countries, some of which offer the bonus of limited employment legislation. Based on the literature, one result may be the mercenary treatment of workers under reincarnated Taylorised regimes, where the complicity of various HRM policies and practices should not go unnoticed.

The way forward?

According to the UK's HSE, organisational culture is the key to managing the OHS problems recorded in the call centre industry. In their most recent investigation of UK call centre working practices, they discuss a number of key OHS hazards in some depth, hoping to provide call centre managers with guidance on how to create humane, healthy and safe working environments for their staff. The revelations include the importance of 'excellent control systems for environmental conditions (because) bacteria and viruses spread more easily in open plan offices', along with 'call handlers are at risk from work-related stress when they are given too much work to do in the time allocated or are not trained how to do the work . . . if call handlers feel they have too much to do, they may not take their breaks . . . call centres should emphasise the importance of rest breaks or changes in activity' (HELA/HSE 2001: 26). The practice of stating the obvious continues with:

> Organisations with positive cultures will have a clear sickness/absence policy and should encourage call handlers to recover fully before returning to work from sick leave in order to protect the health of other employees.
>
> (HELA/HSE 2001: 26)

The fact that the HSE engages in these types of cajoling public relations exercises (as part of the 'Good Business is Good Health' campaign) merely reflects the political boundaries within which it operates, while low levels of full compliance to the DSE regulations (amongst others) suggest the failure of these exercises, as well as the limitations of legislation. For example, one study reports that out of those UK employers who had complied with the DSE regulations, half had complied 'well', while the other half had complied 'a little' (HSE 1998a). And, despite this, HSE inspectors issued only fourteen improvement notices under the DSE Regulations (1992) between 1 January 1993 and 31 March 1999[6] (HSC 2000b). The charade of a 'business-friendly' approach to enforcing OHS legislation appears, therefore, to hold little benefit for call centre workers. This task may instead befall trade-union organisations.

Leading the way forward: trade-union organisations

While UK call centres have been a popular playground for anti-union tactics, the high population of finance-sector organisations (which have a traditionally strong union presence), has meant that in the UK unionisation levels are fairly high. One UK study of six call centres reported unionisation levels as high as 84 per cent (Taylor and Bain 1999). Overall union density figures in UK call centres are reported to be 44 per cent (IDS 2000), which compares to an overall UK national level of 42 per cent in 1998

(Cully *et al.* 1999: 238). The relatively high levels of unionisation in the UK call centre industry may be explained by a combination of factors, including the statutory right to recognition, along with targeted recruitment campaigns in call centres. Perhaps equally influential is the adoption of various exploitative employment practices by some employers who, consequently, may have unwittingly reinstated trade unions in the employment relationship as well as reinvigorating trade unionism. This is evidenced in part by the fact that unions have been successful in recruiting large numbers of workers, including those groups not accustomed to a trade-union culture (for example, young female workers and agency workers), and also by the range of key concessions and improvements won by trade unions to date. As argued by Taylor and Bain (2001: 41), 'call centres are fertile soil for trade union recruitment and organisation, and the intensity with which workers in many locations were being driven would inevitably produce collectively organised responses'. One such collective response is the strike action taken by 4,000 Communication Workers Union (CWU) members on 22 November 1999.

In the first UK strike activity specifically over call centre issues, call centre workers and CWU members in thirty-seven British Telecom (BT) customer service centres across the UK expressed their resistance to exploitative working practices. The dispute opened the way to joint negotiation over an agreement to establish and implement a set of model working practices for the industry. Only a few months later, the media exposed 'black-hole' working conditions at Excel Multimedia sites in Glasgow and Birmingham. Examples of the so-called 'dehumanised work' included targets to handle at least two calls every minute for seven hours a day and the deduction of lost time (for example, toilet breaks) from financial bonuses (IDS 2000). As a result of the bad publicity, Excel Multimedia lost the contract to handle emergency calls and directory enquiries for Cable and Wireless.

The international pervasiveness of unsatisfactory working conditions in the call centre industry is confirmed by a European survey of employment terms and conditions. It reports a number of key problems concerning low pay, enforced working-time flexibility, a high incidence of atypical working, high stress levels resulting from the nature of the work and the working environment, a high incidence of employee monitoring, unrealistic work targets and a variety of health and safety issues – all of which combine to produce a disturbing scene of contemporary work (*European Industrial Relations Review*, 320, 2000b: 15). It is not surprising, therefore, that employment practices and employment relations in European call centres are key policy issues for the European Commission. In many cases, trade unions have acted as a catalyst for improvements in the industry. In addition, some trade unions have seized opportunities for extending the bargaining agenda. As Taylor and Bain (2001: 62) explain, call centre workers want their unions to extend their influence from 'traditional' issues (such as pay) to 'point of production' issues (including those relating to the pace and intensity of work as well as the quality of the working environment). This would involve agreement on reduced targets, an increased number and length of rest breaks, longer call-handling times, higher staffing levels and a greater emphasis on employee development. This approach has been adopted by Swedish trade unions where a fresh focus on working conditions extends beyond the more general issues such as pay and working hours towards coverage of work processes and work organisation (Lindgren and Sederblad 2000).

Conclusions

The call centre industry provides an interesting gateway into a contemporary technology-based workplace and an opportunity to explore contemporary OHS issues. At a macro level, we can gauge the role and influence of government policies and legislation in shaping the OHS landscape in the call centre industry. It appears that while governments and policymakers have been keen to encourage and stimulate employment through call centre growth, a corresponding rise in work-related illnesses and injuries has occurred. While governments have enacted a range of regulations providing workers with rights to safe and healthy workplaces and working practices, the internecine forces of a self-regulatory approach to OHS and the business-friendly approach of the regulatory agencies appear to devalue and deconstruct the intended benefits of the legislation.

In an industry that epitomises late twentieth-century technological developments, examples of the very same exploitative managerial practices of the early factories at the beginning of that century appear to be applied with the same level of managerial vim and enthusiasm at the start of the twenty-first century. In the new 'factory' setting, the assembly lines of manufacturing plants appear in some cases to be replaced by pods of call centre agents constructing the product called 'customer service'. In this setting, 'ye olde worke' intensification strategies are given new life by electronic surveillance and the related monitoring capabilities, while the aspersions cast on management's intentions for technology by Braverman (1974) appear in some cases to be true, evidenced by examples where the technology has been used to emasculate rather than empower workers. It seems that everything has changed but nothing is different – especially in relation to the subjugation of workers' health by profit strategies.

The feelings of *déjà vu* continue when we consider the international division of labour – an established pattern in the behaviour of employers seeking lower costs in production – that is also evident in the call centre industry. While research evidence is somewhat contradictory, it would be naïve to rule out the possibility of opportunistic behaviour from some employers who enthusiastically capitalise upon looser regulatory conditions and lower wage costs in developing countries. For these countries, their experience of the 'information age' may represent the antithesis of Toffler's vision, in which the reality is one where technologies simply reinforce and regenerate patterns of exploitation and subjugation. Without international pressure to harmonise working practices and conditions, as well as workers' rights (with particular regard to the right of representation) throughout the international call centre industry, the expected call centre boom in developing countries understandably generates concerns about employee health.

More encouraging is the research that suggests that some call centres are discarding many of the 'sweatshop' practices in favour of policies resonant with 'high-commitment management'. This comes partly in response to expressions of employee resistance and a level of cognisance over the relationship between intensive working practices and employee costs (for example, absence, turnover) as reported in various research studies, as well as skills shortages in particular areas. As working practices and conditions in the international call centre industry evolve, it will be interesting to note the extent to which OHS improvements occur as a result of reactive 'quick-fix' responses to high turnover and absence, or, alternatively, the result of proactive responses grounded in the recognition of the potent effect of OHS on employee and business performance. Based on the literature, it appears that once again employee health is overlooked when designing and implementing a range of people management policies and practices affecting both work organisation and the working environment. However, a reinvigorated interest in 'safety culture' may hold some promise of brighter prospects for OHS management. This leads us into the next chapter, where we explore HRM and OHS in the international nuclear power industry. In recent years, the industry has embraced the concept of 'safety culture', which attempts to build, develop and sustain the most appropriate attitudes and behaviours that support 'good' practice in OHS. A range of HRM principles, policies and practices are shown to have an important contributory role in achieving this goal, allowing some insight into the potential value of HRM in high quality OHS management.

7 HRM and OHS

Safety culture in the international nuclear power industry

At this stage, it is clear that while HRM-related policies and practices can successfully maximise the human resource contribution, this may incur a significant cost for employees. The deleterious effect of 'human resource maximising policies' was all too apparent in the case studies of airline cabin crews and call centre workers, in which various policies and practices were intimately linked to work intensification and work-related illnesses and injuries. Based on these findings, it could be argued that employee health is a relatively low priority when implementing various HR policies and practices, with short-term economic imperatives trampling over any other contenders for higher positions in the business agenda. To some degree, we have observed how this unfair competition is aided and abetted by policymakers who make decisions on legislative provisions, while the role of enforcement-agency (in)action and the degree of employee (and trade union) involvement also contribute to the fate of OHS. However, when the stakes for 'good' OHS outcomes are higher, for example, where organisations face crippling liability insurance/compensation costs, or where the fragility of public support for certain industries places an imperative on the avoidance of OHS failures (for example the nuclear power industry), OHS might be expected to occupy a higher position in management and HRM agendas. One vehicle for the upward movement of OHS within these agendas is 'safety culture'.

'Safety culture' has been described as the product of the values, attitudes and behaviours of employers and employees in connection with workplace safety. Since HRM and its related practices and policies attempt to influence (and capture) employee attitudes and values, 'safety culture' fits well with the orientation of this text. A broad review of 'safety culture' is possible within the context of the nuclear power industry, where it is a dominant theme (see, for example, Wilpert and Itoigawa 2002; Harvey *et al.* 2002; Harvey *et al.* 2001; Lee 1998; IAEA 1996). Indeed, within this industry 'safety culture' has been heralded as a key predictor of safety performance (Advisory Committee on the Safety of Nuclear Installations 1993).

It is hard to imagine another industry that has the same potentially devastating global impact on health and safety. Thankfully, serious accidents are uncommon although the number of 'incidents' appears to be growing. In the UK, for example, the total number of 'reportable' incidents at nuclear installations in 1999 was fifteen, compared to twelve in 1998 and eleven in 1997 (HSE 2000e). However, on an international basis, the literature suggests both a higher occurrence and more dramatic increases (IAEA 2001a). While in some countries nuclear power is portrayed as a 'sunset industry', its growth and development in other countries (for example, Japan and India) along with the heavy undertaking involved in decommissioning old nuclear installations in the UK and elsewhere, underline the topicality and ongoing importance of the industry within OHS debates. The significance and value of effective health and safety management in the industry finds expression in the promotion of a 'positive safety culture'.

According to the International Atomic Energy Agency (IAEA), a safety culture is an 'assembly of characteristics and attitudes in organisations and individuals which establishes that, as an overriding priority, nuclear safety issues receive the attention warranted by their significance which co-ordinates safety in the industry' (IAEA 1988). Similarly, the Advisory Committee on the Safety of Nuclear Installations (ACSNI) defines a 'positive safety culture' as 'the product of individual and group values, attitudes, perceptions, competencies, and patterns of behaviour that determine the commitment to, and the style and proficiency of, an organisation's health and safety management' (HSC 1993: 23). The obvious parallels with organisational culture and safety culture are summed up by Clarke (2000: 75): 'Like organisational culture, safety culture might be defined as representing the basic values, beliefs and assumptions concerning safety that are embedded in the organisation.'

As such, 'safety culture' is an expression and product of management's and employees' attitudes and values. Notwithstanding a range of concerns regarding the validity of the construct, we consider the literature on 'safety culture' useful in providing further insight into the priority given to OHS in business and HRM agendas. For example, the literature identifies a range of HRM 'essentials' that contribute to an organisation's 'safety culture' including trust and consensus between management and employees, reward strategies, employee involvement, communication and consultation structures and processes, as well as management and employee commitment to safety goals and objectives (Wilpert and Itoigawa 2002; Cheyne *et al.* 1998; Cohen 1997; Smith, Cohen and Cohen 1978; HSE 1989). Indeed, poorly trained staff, along with inadequate or the absence of, regulations, have been implicated in the major international radiation accidents over the past fifty years (Nenot 1998: 437).

Within this industry, the dominant role of regulations and enforcement agents' activities on 'safety culture' and OHS outcomes is a resounding theme. According to one manager at a UK nuclear power installation interviewed by the author, 'without the current tight regulation and monitoring, operators would be less rigorous and more accidents would happen'.

In many Western countries the nuclear power industry is subject to considerably more regulatory controls than other industries, both in terms of the scope of regulations governing their activities (for example, decommissioning plants) and inspection rigour. Also setting this industry apart from others is the extent of 'self-regulation'. In the UK, for example, the boundaries of 'self-regulation' appear to be limited to the formulation and maintenance of safety programmes that have to win the approval of the HSE and its nuclear-plant inspection branch, the Nuclear Installations Inspectorate (NII). This approach endeavours to allow adequate freedom for operators to produce safety programmes that are best suited to their unique circumstances, with the add-on of strict monitoring. However, a dilemma of design is apparent where a culture of blame and a zero-tolerance approach may be propagated by stricter regulation, monitoring and sanction, leading to, for example, cover-ups, faulty reporting and limited opportunities for organisational learning. As a safety manager at a UK nuclear installation told the author, 'How do you manage the hearts and minds of workers when you have a zero tolerance policy which scares the life out of the workforce. In my view this is not the way to go because they will cover up.' In his view, some of this negativity might be neutralised if the conditions of strict regulation and monitoring are complemented with proactive and supportive attitudes and behaviours from employers and employees. Based on the literature, it is these types of attitudes and behaviours that underpin a strong 'safety culture'.

The renewed interest in 'safety culture' also brings to the forefront the criticality of meaningful employee involvement in the management and regulation of workplace safety. This is reminiscent of judgements passed by many commentators since the failed *laissez-faire* experiment in the nineteenth century – that safe and healthy workplaces require meaningful contribution from the workers, who in turn must be given the power to assert and act upon their views and decisions. However, it appears that capital is up to the same old tricks, as safety culture assessments become increasingly captured by management and 'specialists' acting on behalf of employers, which all too often means that workers are kept on the sidelines rather than playing on the field.

This chapter is organised as follows: we begin with a summary of the main criticisms of the 'safety culture' as a valid construct. This is useful in highlighting the complexities of culture and the flaws in current approaches to measuring aspects of 'safety culture'. This is followed by an overview of key industry trends, after which we consider the impact, flaws and potential of 'safety culture' within the industry. A review of the flaws in 'safety culture' is instructive in underlining some of the fallacies associated with HRM, such as the creation of an organisational climate characterised by trust, consensus and openness, as well as the limitations of HRM principles in current economic and political climates.

Safety culture

The importance of organisational culture to corporate success is well documented in the HRM literature, and, more recently, the same importance has been attached to 'safety culture'. According to Cox and Flin (1998: 190), 'safety culture' first came to prominence in the aftermath of the Chernobyl nuclear accident in 1986. Since then, proponents have galloped ahead promoting 'safety culture' as a primary lever in organisational performance and effectiveness, despite uncertainty about the construct's meaning and measurement. Its precociousness is further supported by a front-stage appearance in several UK public inquiry reports following various industrial tragedies. For example, 'poor safety culture' was implicated in the Piper Alpha disaster and the Clapham Junction rail tragedy, while the UK government has proposed that an 'excellent safety culture' in the UK nuclear power industry precludes a Chernobyl-type accident (ministerial statement 1987, quoted in Cox and Flin 1998: 190). The macabre irony is that Soviet nuclear engineers had made a very similar claim six years prior to the Chernobyl accident when they argued that a Three Mile Island-type accident could not occur in the Soviet Union because of qualitative differences in 'safety culture' (Pidgeon 1998: 203).

A plethora of literature on 'safety culture' from industrial bodies such as the Confederation of British Industry (CBI) and regulators such as the HSE has emerged and, to some degree, this has led to a range of 'tick-box' audits. While current mechanisms undoubtedly have some flaws, some of which will be highlighted, these developments are nonetheless important steps towards greater scrutiny being directed at the complex range of factors that influence industrial safety behaviours and outcomes – although it is important to note the reactive, as opposed to proactive, basis of these developments in many cases.

However, there is a considerable danger that 'safety culture' is a fool's gold, in that misguided credence is placed upon its effect on OHS outcomes. Pidgeon (1998: 213) identifies a number of theoretical dilemmas in the 'safety culture' literature. A review of these provides a useful framework for our forthcoming critique of current practice. The first, referred to as the 'basic paradox of safety culture', arises from the central role played by culture in our understanding of the world and organisations, which he argues can serve to illuminate some hazards yet at the same time deflect attention away from others. As Höpfl (1994: 53) explains:

> The pursuit of coherence imposes the appearance of order on a wide range of behaviours and experiences including the discrepant and irrational. Thus, multiple meanings are likely to be glossed over by a privileged interpretation of events ... The point here is that corporate culture change seeks to reinforce corporate norms and in doing so may foreclose on some of the wider interpretations of an event with detrimental consequences for safety.

She goes on to critique the notion of cultural consensus whereby management's attempt to define and categorise aspects of safety create bounded decision zones and may not accommodate multiple meanings, unexpected events or information, leading to the particular problem not being officially recognised as a hazard. In her words, 'The problem for safety management is that it is what is left outside of this "bounded rationality" which is likely to be far more hazardous than those aspects of the system which has been anticipated' (Höpfl 1994: 54).

This connects to Pidgeon's (1998) second dilemma and to popular risk-management strategies. In these we find a collection of anticipated hazards that serve to provide a rational framework for the management of safety. However, as mentioned above, irrational and unexpected events may not be accommodated within such a framework and as a result these may be ignored or misinterpreted. This concern is intensified by efforts to condense those acknowledged aspects of 'safety culture' into simple formulations and quantifications, where it is argued that safety issues are reduced to sanitised rhetoric and a cosmetic exercise that is supported by artefacts such as manuals, quality (and safety) assurance awards and policies (Höpfl 1994). The danger of over-simplified and superficial analysis of 'safety culture' may mean that the underlying organisational and social factors, such as poor communication between management and front-line workers, are underestimated and even ignored. According to Reason (1990), if more attention were directed towards identifying and remedying these types of underlying organisational or 'latent' failures, greater safety benefits would be achieved. Such arguments find support in Sagan's (1993) case study of the nuclear power industry. The author reports that on the surface, many of the facets of a 'good safety culture' were evident, such as priority given to safety by senior politicians and decision makers, adequate resources devoted to safety and the opportunity to learn from mistakes. The presence of these did not, however, prevent a number of near-miss nuclear weapons launches and detonations. This was explained in terms of attitudes and behaviours that led to organisational blindness, to errors/risks and the inhibition of interrogative thinking.

Other problems are apparent in terms of the extent to which culture and its associated values are evenly spread across any given organisation. This position ignores the likely possibility that different groups (for example, management and shopfloor workers) may hold different sets of cultural values (such as those relating to safety). The presence of group subcultures within organisational culture is underlined by a number of commentators (for example, Beck and Woolfson 1999; Clarke 1999; Hofstede 1991), and is supported by Harvey *et al.*'s (2002) study of two UK nuclear plants. The authors identified three distinct groupings leading them to conclude that two or more safety cultures were present in the organisations. As such, the difficulties involved in identifying and defining organisational subcultures present considerable challenges not only for the development of a sound theoretical framework for 'safety culture', but also for a better understanding of rule violations and perceptions of 'acceptable risks' – issues that align with the third dilemma identified by Pidgeon (1998).

Pidgeon (1998: 20) also notes how the question of politics and power is curiously absent in safety culture debates and emerging theoretical models. Equally, these factors may exert a strong influence on the extent to which organisational learning will take place. For example, parochial interests may lead to a range of safety misdemeanours such as secrecy and under-reporting (Rijpma 1996). The point here is that organisational politics, particularly in the face of high levels of external accountability, may inhibit and even preclude an awareness of incumbent and emergent hazards, as well as the capability to learn from mistakes (Sagan 1993, Pidgeon 1998). As Pidgeon and O'Leary (1994) have argued, the essence of safety culture will be shaped by the processes of organisational intelligence gathering and interpretation, which in turn can enhance, or limit, the organisation's capacity for corporate myopia.

Overall, many of the criticisms relating to the theoretical underpinnings of safety culture are due in part to limited data, which prohibits the testing of the reliability, validity and utility of existing definitions and measures. Some of the problems in definition and measurement emanate from unclear terminology, such as the interchangeable use of 'safety culture' and 'safety climate'. Such differentiation is necessary since the latter refers to attitudes, perceptions and beliefs, while the former is more complex, reflecting values and norms that are mirrored by safety management practices (Harvey *et al.* 2002: 19). However, the literature notes a degree of symbiosis between the two. Cheyne *et al.* (1998: 256) argue that a 'good' safety culture will be promoted and maintained by a 'good' safety climate and vice versa. Our interest is not in the intricacies of the safety culture versus climate debates, but directed more towards the usefulness of 'safety culture' in enhancing the profile of OHS in business and HRM agendas. The opportunities for this are, however, likely to be inhibited by some, or all, of the flaws highlighted above. Before considering 'safety culture' in operation, let us first review some key industry trends in the international nuclear power industry in order to appreciate the range of economic, political and social influences on OHS policy, practice and outcomes.

Industry trends

According to the International Atomic Energy Agency (IAEA), at the end of 2000, there were 438 nuclear power reactors operating around the world, 651 nuclear power reactors (of these 284 are in operation) and 250 fuel-cycle plants, including uranium mills and plants that convert, enrich, store and reprocess nuclear material (IAEA 2001a). In addition, the construction of three new nuclear reactors started in 2000 – one in China and two in Japan – bringing the number of nuclear reactors under construction at time of writing to 31.

Table 7.1 Countries' reliance on nuclear power in 2000

Country	Percentage of reliance on nuclear power in 2000
France	76.4
Lithuania	73.7
Belgium	56.8
Slovak Republic	53.4
Ukraine	47.3
Bulgaria	45.0
Hungary	42.2
Republic of Korea	40.7
Sweden	39.0
Switzerland	38.2

Source: IAEA (2001a).

Nuclear power provides about 16 per cent of global electricity, with about 83 per cent of nuclear capacity concentrated in industrialised countries. The countries most reliant on nuclear power in 2000 are shown in Table 7.1.

In North America, 118 reactors supply about 20 per cent of electricity in the USA and 12 per cent in Canada. In these countries and in western Europe, the number of operating reactors is declining, albeit only slightly. Only in the Middle East, Far East and South Asia, which in 2000 had in total 94 reactors, are there clear plans for the expansion of nuclear power, particularly in China, India, the Republic of Korea and Japan (IAEA 2001a). Despite offering 'a cost effective, clean and safe method of energy production', the nuclear power industry has struggled against a swell of public distrust and negativity.

At one international conference, various countries shared their concerns over securing public support for nuclear power. The UK representatives presented findings of the UK's Nuclear Installations Inspectorate's (NII) 'Public Opinion Survey' (2001),[1] which offers a gauge on the level of public concern about nuclear-related issues. Given that the UK is allegedly one of the safest nuclear countries, it is interesting to note that two out of five respondents held a negative attitude towards the nuclear power industry, with only one in five reporting a positive attitude. The top three drawbacks of nuclear power were reported as nuclear waste, the risk of accidents and possible health risks. Over half of respondents could not name a single benefit (NII 2001: 157). Over half of respondents felt the most worrying aspects of nuclear power was the possibility of a major accident and release of deadly or dangerous radioactivity, while 42 per cent were concerned that over time a nuclear plant might be adversely affecting the health of the local population. In addition, a high proportion of the sample disagreed that people can safely live near a nuclear site, that nuclear waste can be stored safely and that nuclear energy is less harmful to the environment than other sources (NII 2001: 158). Indeed, respondents' level of concern over nuclear waste was second only to their concern about crime (NII 2001: 159).

At the same conference where these findings were presented, Mr Matsuura, Chairman of the Nuclear Safety Commission in Japan, discussed the efforts being made to alleviate public distrust and anxiety caused by incidents and scandals in his country's nuclear power industry, not least the devastating Tokaimura accident of September 1999. Mr Matsuura reported the results of public opinion surveys following the Tokaimura accident, noting that 'there has been a steep increase in anxiety about nuclear power', although 70 per cent of the sample believed that 'atomic energy is valuable' (OECD 2001: 27). Restoring public confidence in nuclear power, according to Mr Matsuura, will be dependent on eliminating accidents and incidents. To some degree, this will be contingent upon the adequacy of regulations and activities of the country's nuclear safety commission. These sentiments are summed up by another conference delegate:

> Trust means that you have to invest in an independent regulator, with an open attitude and with the capacity and competence to review the safety assessment done by industry. You have also to invest in the regulator's ability to act as the people's expert in stretching industry ... You also have to invest in a legal framework which clearly states the responsibility between industry and regulatory bodies.
>
> (J. Melin, Director General of the Swedish Nuclear Power Inspectorate, OECD 2001: 51)

The high priority placed on minimising OHS failures brings together the combined forces and influences of comprehensive legislation, empowered and invigorated enforcement agency activities and proactive employer and employee attitudes towards OHS (albeit this occurs at varying degrees in different countries). In addition, a heavy reliance on employee input into the day-to-day management of workplace safety appears to be highly valued. Combined, all of these elements provide a possible framework for a paradigm for 'good OHS practice' (and 'safety culture'). However, the literature suggests that there is some distance between policy and practice in the international nuclear community.

One of the main inhibiting factors in achieving more effective regulation and monitoring in the international nuclear power industry is the growing skills crisis. According to the Nuclear Energy Agency (NEA) (2000), there is some doubt over the long-term ability to preserve safety competence within the nuclear power industry and its regulators because of a declining number of available skilled personnel. In a presentation to the NEA conference, the UK NII reports that over the next ten years, attrition rates for staff are expected to be as high as 40 per cent. In their words:

Quality organisations require well educated, well trained and well motivated staff. In some countries, national R&D programmes are being reduced to such a point that forming an independent regulatory position might be in jeopardy. If a significant problem occurred over the next ten years, there might not be sufficient knowledge and capability to deal with it in a timely manner if the current trend continues.

(Nuclear Energy Agency (NEA) 2000: 11)

However, the picture is mixed across countries, highlighting the interplay of social, political and economic views about the place of nuclear power within national energy policy. Recruiting and retaining skilled staff is fraught with difficulties in countries experiencing a nuclear power slow-down. In many countries the number of undergraduate programmes in this field (in particular, safety research) have declined, and as programmes close, there is less research support available (NEA 1999). In addition, educators and industry specialists are ageing and, as they retire, there is an acute lack of competent and capable people to step into their shoes.

It is clear then that concerns over human resource planning and skills shortages in the international nuclear power industry present major challenges for securing future competence in the industry. Given that nuclear power-plant accidents already occur globally at a rate of approximately three per year (Department of the Environment 1997: 5), it is not surprising to find these issues sitting beside 'safety culture' at the top of the IAEA's agenda, particularly in relation to countries where regulatory frameworks and enforcement are less rigorous than in, for example, the UK. Indeed, according to one UK academic specialising in the nuclear power industry who was interviewed by the author, the perilous nature of nuclear installations in some eastern European countries (for example Bulgaria) represents a major stumbling block to their ascendance into the European Union. This brings the regulatory framework into focus.

Regulating the industry

The IAEA, a UN agency, is based in Vienna and has 132 member states. It serves as the world's intergovernmental forum for scientific and technical co-operation in peaceful uses of nuclear energy. One role of the IAEA is to provide safety-related technical assistance in the form of expert services, equipment and training. In 1998, the IAEA reported a total of some 180 national, regional and inter-regional projects, at a total cost of around US$15.8 million, of which 45 per cent was devoted to nuclear safety and 55 per cent to radiation and waste safety. In addition, major extra-budgetary programmes on the safety of nuclear power plants in eastern Europe, the former Soviet Union, south-east Asia and the Far East were underway (IAEA 1998). Following the terrorist attacks on the USA on 11 September 2001, the IAEA member states adopted a resolution on the physical protection of nuclear materials and facilities to strengthen programmes related to the pre-vention of nuclear-related terrorism. According to the agency's president, Mohammed El Baradei: 'September 11 presents us with a clear and present danger and a global threat that requires global action'.

In the short term, the IAEA estimates that at least US$30m–50m annu-ally will be needed to strengthen and expand its programmes to meet the terrorist threat. Without detracting from the considerable threat that terror-ists present to global nuclear safety, it is important to note that nuclear power presents a continual threat to global safety given the fact that, between 1995 and 1997, forty-five accidents were recorded by the IAEA (thirty in 1997; thirteen in 1996; two in 1995) (IAEA 1998).

Rather ominously, the IAEA's safety-assessment team (Assessment of Safety Significant Events Team) undertook fourteen investigations or, in their terminology, 'missions' in 1997 in eleven different countries, includ-ing the UK, Canada, Egypt, Spain, Sweden, Ukraine, Kazakhstan and Lithuania, and reported that: 'Safety problems related to personnel profi-ciency and adequacy of procedures are still of concern, but appear to be decreasing' (IAEA 1998). In addition, the IAEA's International Regulatory Review Team (IRRT) service will, on request, not by decree, review the effectiveness of the relevant regulatory bodies in countries that 'express an interest in this service' (IAEA 1998). A case for increasing the scope and breadth of the IAEA's activities is given some support when we consider that the Japanese nuclear regulatory agency's last inspection of the Tokaimura plant was seven years prior to the accident in September 1999.

However, a counter argument for increasing the inspection and sanction-ing powers of the IAEA is that a shift from its current 'non-threatening' status could inhibit the propensity of nuclear power companies and agen-cies to report 'unusual' events in nuclear power plants, thus damaging the current tentative openness in the international community. Presently, information on 'unusual' events and incidents is collected and analysed by the IAEA through the Incident Reporting System (IRS). The system desig-nates to each member country a national IRS co-ordinator who reports 'safety-significant' events. Following such an event, a detailed analysis and report is communicated to all member countries 'for professional use'. Public access to the database is denied. In a statement on the website, the IAEA report that 'virtually every country with a nuclear power programme participates in the IRS'. No details are provided on which countries are the exceptions. They also state without any sign of discomfiture that 'there are now over 2400 reports on the system, with new ones added at about a rate of 100 per year' (IAEA 2001b). It is important to note that it is only the accidents that are deemed to be of 'international interest' by the national co-ordinator that are reported, leaving some uncertainty about the actual number of safety-critical events that actually occur every year. The opaque transparency in the industry is further highlighted by the potentially subjective nature of the decisions about which incidents are deemed 'safety significant'. On a more reassuring note, there is wide evidence of positive action from international regulatory and enforcement agencies to develop a cohesive international network of safer plant operation, highlighting the role and indeed absolute prerequisite of weighty external intervention in OHS.

Recent investigations of international nuclear accidents are useful in exploring 'safety culture' in the industry, not least to review four of the main criticisms levelled against the construct at the beginning of the chapter. First, the proposition that 'safety culture' is not adequately captured by cultural artefacts such as manuals and declared rhetoric could be supported by evidence of a disjuncture between policy and practice; second, criticisms of managerially-defined rational frameworks for installing and measuring 'safety culture' would be endorsed by evidence of outlying hazards that present a potentially greater danger than those hazards identified in the system; third, evidence of different sets of cultural values and attitudes within a presupposed single 'safety culture' would lend support to criticisms levelled against cultural consensus in organisations; fourth, examples of power and politics within organisations affecting safety culture norms and organisational learning would add credence to calls for the inclusion of these factors in the theoretical development of the construct – all of which would go some way to disqualify many of the superficial and simplistic approaches applied to the management of workplace safety at both governmental and organisational levels.

The Tokaimura nuclear accident

> The accident at the JCO [Company's] nuclear fuel processing facility at Tokaimura seems to have resulted primarily from human error and serious breaches of safety principles, which together led to a criticality event ... It will most probably have implications for the regulatory regime and safety procedures and safety culture at the JCO facility.
>
> (IAEA 1999: 33)

On 30 September 1999 at 10.35 am local time, an inadvertent nuclear chain reaction occurred in Japan's Tokaimura nuclear power plant 120 miles northeast of Tokyo. Despite assurances made when the plant was licensed in 1983 that 'critical fission chain reactions could not occur', the chain reaction expelled intense heat and radiation for 18 hours before it was eventually stopped. The IAEA (1999) reported that in addition to the three affected workers, fifty-six people at the JCO facility were confirmed to have been exposed to gamma and neutron irradiation, plus a further seven workers who were working on a construction site nearby. The three emergency service workers who transported the three JCO workers to hospital were also confirmed to have been exposed to gamma and neutron irradiation. Following a delay of nearly an hour in informing the local authorities of the accident, police blocked off a 250 m radius around the plant and after a further delay of nearly 4.5 hours, police evacuated some 161 local residents in a 350 m radius of the plant and some 310,000 people were advised to stay indoors for about 18 hours as a precautionary measure (IAEA 1999: 1). Less than 24 hours after the event, the Japanese government issued an all-clear (around 7 am the following morning) (Dolley 1999).

Accounts of the accident reported that it resulted from a direct violation of national nuclear power safety regulations. However, the action that led to the accident – rated (somewhat controversially) 4 on the International Nuclear Event Scale (INES)[2] – was a standard operating procedure practised for over two years that was codified in a secret company manual but was never shown to government safety regulators (Dolley 1999). As Steven Dolley, Research Director for the Nuclear Control Institute, explains:

> The accident began when workers were converting enriched uranium into oxide powder for use in preparing fuel for the fast breeder reactor. One of the injured workers reported that some 16 kilograms of uranium solution had been poured into the precipitation tank, nearly eight times more than its criticality safety limit of 2.4 kilograms. Workers reported seeing a blue flash and then started to feel ill. The area was reported to be wrapped in a haze of blue smoke. The criticality continued for about 18 hours until the water that was moderating the flow of neutrons and allowing the chain reaction to continue was drained and the tank was flooded with boron, a neutron absorber.
>
> (Dolley 1999: 1)

The IAEA (1999: 11) provide a similar account of the actions that led to the accident, in which unsafe working practices are attributed as the cause:

> The work procedure was modified in November 1996, without permission [from the regulatory authorities] . . . this new procedure had been followed several times before this accident occurred. Furthermore, when the criticality event occurred, homogenisation of uranium oxide was being performed by mechanical stirring in the precipitation tank instead of in the mass control equipment . . . This means of homogenisation in the precipitation tank is not even described in the revised procedure and was a further deviation from the approved procedure.

The IAEA report goes on to describe how the situation was brought under control by injecting aqueous boric acid into the precipitation tank. The boric acid feeding operation did not begin until 8.19 am the following day for a number of reasons, one of which was that there were no stores of boric acid at the JCO site (IAEA 1999: 12).

According to the Japanese government, no radioactive fall-out occurred, despite gaseous releases into the environment occurring through the building's ventilation system (IAEA 1999: 23). However, according to one source, radiation levels near the plant were up to 15,000 times above normal background levels (Dolley 1999). The discrepancy in the accounts is not surprising since the government agency, the Mito Atomic Energy Office of the Science and Technology Agency (STA), failed to take any measurements until almost three and a half hours after the accident (IAEA 1999: 27). While the government reported that background radiation levels returned to a 'normal' level within 24 hours of the accident, measurements of radiation levels taken three days after the accident on a public road 30 m from the nuclear power plant revealed levels of radioactivity five times the 'normal' background levels (Dolley 1999).

According to the IAEA, a survey conducted on 1 October of the ground surface within a 700 m radius showed no radioactivity contamination that could be associated with the accident. Nonetheless, Greenpeace has called for long-term health monitoring programmes for the people living in the immediate vicinity of the plant, as well as workers in the fields and people passing through the area during the 4.5 hour delay from the accident occurring to the eventual evacuation and closure of the area surrounding the nuclear power plant. Greenpeace also argues that since the 350 m radius zone was never completely evacuated and that the area affected was more likely to be at a radius of least 500 m from the site of the accident, many people were exposed to the full 18 hours of neutron radiation – the health effects of which will not begin to show for at least ten years (for example, blood cancers and other longer-term induced cancers). The government response was to issue 1 million copies of a pamphlet on or around 15 October 1999 to the 74,633 residents in the Ibaraki Prefecture affected by the accident.

The fate of the three workers involved in the accident – an accident caused by unsafe working practices, infringement of the law and the failure of the regulatory agencies to detect irregularities – is described below:

> One and half hours after the initial exposure, patient A was conscious but was vomiting and had diarrhoea and fever. It was reported that patient A had a markedly reduced lymphocyte count and marked hypocellular bone marrow (severely damaged bone marrow).
>
> Patient B also had a markedly reduced lymphocyte count and marked hypocellular bone marrow ... the skin of his face, neck, upper thorax and right arm was reddened. The right hand and forearm were reported to have been diffusely swollen and very painful.
>
> Patient C was almost asymptomatic after exposure, with a moderately reduced lymphocyte count and hypocellular bone marrow ... although he would be expected to be subject to an increased risk of incurring cancer or leukaemia at a later date.
>
> (IAEA 1999: 31)

Patient A (name not provided) died two months later from his injuries, while Patient B (Mr Masato Shinohara) died in April 2000 of multiple organ failure following exposure to about 7 sieverts (Sv) of radiation. This is equivalent to around 700 times the recommended maximum exposure for a nuclear-plant worker (IAEA 1999). I was unable to locate further information about the fate of the third victim.

On 1 October 1999, the IAEA Secretariat offered assistance to the Japanese authorities in responding to the accident. This offer was initially declined because the Japanese authorities believed that assistance was not required, but on 13 October 1999, three experts from the IAEA Secretariat (specialising in the nuclear fuel cycle and its regulation, emergency response and accident consequence assessment, and environmental monitoring and dosimetry) began a four-day fact-finding mission in Tokaimura. The intervention of the IAEA is justifiable given the continued support for Japan's nuclear development from Britain, France, Australia, Canada and the USA amongst others, particularly in the expanding area of plutonium and enriched uranium breeder fuels even though these are more susceptible to criticality accidents than conventional power-reactor fuels. Following the JCO accident, regulations were strengthened and special measures for nuclear emergencies introduced along with the installation of safety inspectors and disaster prevention officers at nuclear installations. The name of their paymaster is not specified, but nevertheless the fact that these crucial posts were only created after the accident is a testament to the typically reactive approach to health and safety management so commonly observed across international industrial landscapes.

The Japanese experience suggests that safety in the nuclear power industry requires greater regulatory intervention, independent and empowered enforcement agency activities, as well as a far higher priority being awarded to safety at an operational and strategic level. However, a variety of factors continue to run counter to these goals, notably free market logic. According to the NEA (1999: 14):

> Some companies are claiming that electricity market liberalisation gives rise to better standards of safety ... Paradoxically, preliminary signs show that electricity market deregulation may require a stronger and more effective nuclear regulator.

Are we surprised? The cultivation of regulatory environments where failure to abide by safety and other standards is not only possible but may actually deliver an economic advantage, is never more nonsensical and unsettling than in the nuclear power industry. According to some commentators, this accident marks an increasing trend of deteriorating quality control at Japanese companies in pursuit of cost savings. Investigations of the accident found that officials at JCO skipped routine safety procedures to save money (*Financial Times*, 28 July 2000). That, however, turned out to be a false economy, given reports that JCO would be paying the equivalent of £78 million in compensation to settle 6,875 claims following the Tokaimura disaster (*Guardian*, 5 September 2000), while in October 2000, six officials at JCO were arrested on suspicion of professional negligence based on the inquiry's findings that there was an absence of safety measures and a lack of training (*Financial Times*, 12 October 2000).

The complexity of safety culture is well illustrated by this tragedy, where during an incubation period of well over two years an unsafe working practice was routinely followed and concealed from the regulator by way of surface artefacts such as safety manuals and declared rhetoric. In operating the plant, the absence of safety measures and inadequate training provide some indication of management's commitment to safety culture, while the rule violation that led to the accident highlights managerial prerogative and power inequalities, particularly in terms of workers' apparent 'acceptance of risk'. The government agencies' and the company's inadequate, and by some accounts 'cover-up', responses in the aftermath of the accident may be indicative of caveats within a blame culture. In this example, the fear of sanction may explain the release of, by some accounts, incomplete information, which in turn provides a limited scope for learning and improvement both in the organisation and across the international arena. In a national culture where compliance and obedience is firmly embedded, the Tokaimura accident sends tremors throughout the international arena and brings to the fore Pidgeon's (1998) question as to

> whether organisations can, through a culture of safety or reliability, effectively monitor their own or others' performance and learn from the lessons of the past in ways that can support safety, or whether garbage-can politics of organisational life subvert or suppress what few true learning opportunities there are, particularly when the stakes are high and the intrinsic ambiguity and uncertainty of corporate events renders hindsight a fallible and contested guide.

The Chernobyl nuclear accident

The Chernobyl accident, which occurred on 26 April 1986, is also useful in illustrating some of the dilemmas in safety culture. Analyses of the accident identified administrative and individual attitudes to safety as primary causal factors (Turner, Pidgeon, Blockley and Toft 1989, Pidgeon 1998), again underlining the importance of adequate communication, information, training and meaningful employee involvement in developing and maintaining a strong and effective 'safety culture'. The accident resulted in at least 300,000 people being exposed to radiation dosages of 100–500 mSv. By way of a measure of the enormity of this exposure, under current EU regulations[3], 20 mSv is the maximum recommended annual dose for a nuclear plant worker. The clinically observed effects of exposed individuals included the symptoms of acute radiation syndrome, such as gastrointestinal damage, skin lesions along with thyroid cancer, leukaemia and various psychological effects (for example, depression, anxiety) (IAEA 1996: 3). In addition, a surge in thyroid cancer incidence was observed in children who were born before or within six months of the accident, with very young children being in the highest risk group. Indeed, in its 2000 Report to the UN General Assembly, the United Nations Scientific Committee on the Effects of Atomic Radiation (UNSCEAR) reported that about 1,800 people who were children at the time of the accident had contracted thyroid cancer (IAEA 2001b). The latency period between exposure and diagnosis of thyroid cancer appears to be about four years. The IAEA reports that:

> The extent of the future incidence of thyroid cancers as a result of the Chernobyl accident is very difficult to predict. There remain uncertainties in dose estimates and, although it is not certain that the present increase in the incidence will be sustained in the future, it will most probably persistent for several decades. If the current high relative risk is sustained, there would be a large increase over the coming decades in the incidence of thyroid carcinoma in adults who received high radiation doses as children.
>
> (IAEA 1996: 4)

The above commentary is from the agency that co-ordinates nuclear safety at a global level. The expressed level of uncertainty is rather disconcerting given our unequivocal reliance on this agency to protect global citizens' health. Strenuous efforts are, however, being made to learn from past mistakes. In June 2001, fifteen years after the disaster, scientists and experts gathered at the third international conference for the 'health effects of the Chernobyl accident' to discuss and communicate research following the event, where various prerequisites for a strong and effective 'safety culture' featured prominently.

Countries with better safety records may offer some insight into how to build and sustain a strong 'safety culture'. The UK nuclear power industry has, to date, a relatively good safety record. According to one industry commentator interviewed by the author, 'if there was an international hierarchy of safe nuclear installations, the UK would be in the top three'. However, an array of management failures in the UK nuclear power industry points to more weaknesses in 'safety culture' and, moreover, gives rise to concern about just what is going on further down the international safety hierarchy.

The UK nuclear power industry

In the UK, the HSE, through its NII, regulates safety at nuclear sites under the licensing regime provided for in the Nuclear Installations Act (1965), which complements broader legislation governing workplaces in the UK (for example the HASAWA 1974). The UK Department of the Environment (DOE) (1997: 5) reports that 150 nuclear accidents were recorded, of which thirty-eight resulted in off-site releases. While many did not involved the release of radioactivity or damage to the core, the fact is that accidents (safety failures) occur on a regular basis (three per year). The DOE (1997: 5) reports that accidents that involve the release of radioactive material into the environment occur, on average, every two years. In addition, it is noted that trends up to, and until, 1997 showed a small decline in the number of reported incidents in nuclear facilities, while the number of accidents involving off-site releases has remained approximately constant since the 1960s (DOE 1997: 5).

One such incident occurred on 7 May 1998 at the UK Atomic Energy Authority's (UKAEA) Dounreay nuclear facility, Scotland, when excavation work damaged a buried electrical cable causing an extensive loss of electrical supply causing the deactivation of the visual and audible electrical protection systems (for example, radiation monitoring, criticality alarm systems, radioactive waste discharge monitoring systems). Following this incident, an audit of health and safety was initiated on behalf of the HSE by the Chief Inspector of NII. The audit yielded a total of 143 recommendations under six main themes: Safety Management Systems, Safety Culture, Management and Organisation, Human Resources and Training, Safety Cases and Operational Strategy, with management being particularly criticised for prolonging archaic working practices and adopting a bungled, low-cost approach when changing over to mixed (uranium and plutonium) oxide fuel (MOX) reprocessing (HSE 1998b). In fact, the inspection team were so dissatisfied that they halted operations. The resultant recommendation stated that 'no further processing or reprocessing operations [should take place] until a safety case for those operations has been produced and assessed as adequate' (HSE 1998b: 135). One can only wonder about the number of violations that have occurred in the facility prior to the 1998 audit, given the myriad of weaknesses identified by the audit team in safety management systems.

The story of UKAEA is illustrative of safety management being excluded in strategic planning and management, particularly during organisational restructuring following privatisation in 1994. Set up under government ownership in 1954, the company was split up into three business groups, two of which were sold to the private sector in 1994. The restructuring exercise left some 2,000 workers in the government-owned part, whose main mission was to decommission the redundant installations. Following the government's decision in 1988 to close the facility's Prototype Fast Reactor[4] in 1994, staff numbers fell from 13,600 to 8,300 between 1988 and 1993. The change in emphasis and the staff reductions significantly reduced the company's technical and managerial bases, leading to a heavy reliance on contractors, including those to whom it had just sold parts of its business (HSE 1998b: v). As the HSE notes:

Organisational changes made within UKAEA over the last four years have so weakened the management and technical base at Dounreay that it is not in a good position to tackle its principal mission, which is the decommissioning of the site ... UKAEA must also give urgent attention to other redundant facilities which have in many cases been left severely contaminated with radioactivity and with little attempt made to clean them in preparation for decommissioning ... We suspect that UKAEA has been operating plants without clear knowledge of some of the risks.

(HSE 1998b: vi–vii)

It was not until 1990, after 36 years of self-regulation on nuclear safety matters, that UKAEA was brought into the regulatory regime of the Nuclear Installations Act (1965), and it was not long after that the NII voiced its concerns about the management of Dounreay and the possible impact of privatisation. However, a full audit of management systems at Dounreay was not conducted until the accident of 7 May 1998 – producing prolific evidence of substandard management practices in possibly the world's most perilous industry.

In the HSE's (1998b) view, a number of factors destabilised the safety culture at Dounreay. In particular, the large-scale changes in the organisation and uncertainty over future employment prospects expressed by the workforce. This created a sense of low morale amongst the workforce, which in turn was thought to adversely affect the safety culture in the organisation. However, the positive effect of safety working parties and other forms of communication were clearly acknowledged by the audit team. Safety culture is also affected by the range of skills and expertise held by the workforce – a problem area for UKAEA who were (and still are) experiencing significant problems in recruiting appropriate technical, managerial and supervisory staff with suitable qualifications and experience, with particular regard to safety-related posts. Perhaps these difficulties explain why it was not until April 1998 that UKAEA appointed a Head of Site, who was responsible for safety and for fulfilling the requirements of the nuclear site licence. Up until then UKAEA had been organised on a functional basis and, as a result, managerial responsibility for safety on each site was not vested in a single individual, with the responsibility for tasks including emergency arrangements across the site lying with various managers at different sites. As the HSE noted, such a complex system creates blurred boundaries of responsibility, which is wholly incompatible with effective safety management systems.

A further incongruity identified by the audit team was the safety performance index, which paid a financial bonus (1–2 per cent of salary) if a certain safety target is achieved or improved upon. Targets were based on the number of reported incidents and their corresponding severity. Can you see the problem? Just look for the flashing neon lights that read 'licence for under-reporting'. In order to improve safety, it has become clear that openness is the key to risk identification. A bonus system that penalises workers for such openness is unintelligible and completely inappropriate (in any industry, never mind one of the highest-risk industries). Unsurprisingly, the HSE recommended that the award scheme be scrapped and in its place they suggested a scheme that emphasises the positive factors such as the attainment of specific standards of performance, for example, the number of safety cases reviewed, actions cleared, and the completion of safety tours (HSE 1998b: 23). The audit report concludes that:

> It is evident that UKAEA still needs to invest considerable effort, time, and resource into bringing itself up to the standards of a modern nuclear licensee ... In addressing this, UKAEA needs to set itself high standards and ambitious targets rather than attempting to do the minimum necessary to get by.
>
> (HSE 1998b: vii)

Three years on, UKAEA are still working towards meeting all of the 143 recommendations,[5] providing some indication of the depth and breadth of remedial action and restructuring that was required in the company. This example is clearly useful for illustrating the complexity of safety culture and health and safety management. Any attempt to gauge or, even more ambitiously, measure 'safety culture' at this and any other facility must clearly take into account the influence and interplay between various organisational, managerial and social factors that goes well beyond any simple rationalisations or a solely managerial perspective on what can be regarded and defined as 'safe' or 'adequate'. The interdependencies between various areas of communication, training and reward management, and OHS are also highly evident. To date, it appears that this knowledge is being integrated into remedial action within the plant. While improvements to the site operation are welcome, it is nonetheless important to note that these were driven by external intervention bolstered by heavy sanctions, and that the changes took place only after a potentially serious incident occurred. On a more positive note, Laurence Williams, HSE Director of Nuclear Safety and HM Chief of Nuclear Installations, recently commented:

> I have seen considerable progress at Dounreay over the past three years and this is a credit to everyone concerned, particularly the staff at Dounreay. The Consents we have granted over the last few months to allow the restart of a number of plants are an indication of our growing confidence that safety at Dounreay is improving.
>
> (HSE press release, 22 January 2002)

Conclusions

In a recent paper, Frisch and Gros (2001) attempt to promote an optimistic and positive perspective on the continuous process of safety improvement in nuclear power plants. The authors note that the impetus for improved safety standards in the industry is derived from the goal to increase the acceptance of nuclear technologies, while at the same time they acknowledge that the most significant improvements, or at least those articulated in regulations aimed at raising safety standards, are the product of some of the world's most serious and devastating nuclear accidents. This, to some extent, depicts a scenario of safety advances being based on 'trial and error' in the nuclear power industry, where the mitigation of risk is not the outcome of meticulous advances in safety case analysis, but rather through accident analysis. Learning from past mistakes is of course important, but waiting for accidents to happen before identifying risks and making the necessary improvements is clearly unacceptable. A shift to an open, interrogative and proactive approach to identifying safety risks and unsafe work behaviours is the most effective route to improved OHS outcomes, not least to address widespread under-reporting of safety risks – a practice that creates the false image of a 'safe' organisation, while also concealing the common chain of events that tend to build up prior to an industrial tragedy (for example, Tokaimura, Piper Alpha, Bhopal). As Cox and Flin (1998: 195) argue:

> The absence of an accident to date, [however,] does not prove that this is a 'safe' organisation or even one which has a 'better' safety culture . . . In fact, Sagan (1993) points out that many of these so-called safety organisations have a track record of concealed accidents and safety breaches.

The synergies between legislation, enforcement-agency activities, management and employee attitudes, workplace democracy, community action and public opinion all resonate from this industry example of health and safety management. The correct combination and concentration of the elements in the OHS equation for good practice and outcomes is still not resolved in the international nuclear power industry, but given the international interest in achieving this, one can remain hopeful that continued improvement towards these ends will be made and that the resultant model will be extended to the wider industrial landscape. A cautionary note is, however, necessary with regard to emerging narrow conceptions of 'safety culture' across the industrial spectrum, as manifest in consultancy-driven tick-box questionnaires. A clear danger is that any 'quick-fix', superficial approaches of safety culture dilettantes will fail to capture the complex organisational and social processes involved in OHS and, as such, effect little influence on the rising trend of industrial accidents or, indeed, redress power imbalances in the workplace.

A final point is that many of the idealistic notions associated with HRM, such as consensus, openness, trust and communication, are engrained in 'safety culture', suggesting in turn that HRM principles and policies could, after all, provide a vehicle for elevating the importance of, and the priority given to, OHS. However, the parameters set by political and economic pressures may mean that many of these venerable principles are discarded in favour of something quite different. As we have seen, the prerogative afforded to employers within current political and economic climates finds some nuclear power operators routinely failing to comply with safety regulations, whilst giving little credence to the principles of communication, training and employee involvement in health and safety management. With considerable room for manoeuvre and not forgetting competitive pressures for those countries with privatised nuclear facilities, why should it be assumed that nuclear power plant operators are a different breed from any other profit-motivated employer?

8   The verdict

Our starting position was a review of some of the assumptions contained in the HRM literature. Based on the many flaws identified in these assumptions, we set the question: 'To what extent is HRM capable of providing the optimal conditions for "good" OHS policy, practice and outcomes?' A key sticking point was the suggestion that HRM is inherently unethical on a number of counts, most of which were based in neo-classical and neo-liberal economic theories. In Chapter Two, many of the flaws of free market logic were exposed in relation to their bearing on employee rights and protection. Almost grasping at straws, we suggested that HRM's basic congruence with prevailing economic theory did not necessarily lead to negative consequences for OHS. Then followed the range of economic rationalisations and 'good business cases' for the responsible and ethical management of health and safety. Indeed, in later chapters, some success stories were relayed about how HRM principles had been successfully applied to HSM, particularly those relating to TQM. A 'continuous-improvement' approach seemed to appear most prominently in the 'high-risk' industries, although, as noted, often as a result of a serious accident. A central feature of these programmes was the high level of employee involvement and participation in day-to-day health and safety management. However, while offering a range of good ideas for practitioners seeking to enhance their organisation's approach to HSM, the undeniable vulnerability of these programmes was apparent given that their very existence depended on a benevolent management.

At this point, the inherently political nature of OHS took centre stage. Past and present, conflicts of interest are central to the employment relationship and, as we noted, to most human relationships. And while the means used to resolve different or opposing interests are a good indicator of the nature of 'democracy' in any society, policymakers' approach to democratic employment relations appeared remarkably similar to that of a dog owner. For example, employees around the world in First-, Second- and Third-World countries appeared to be, albeit to quite different degrees, kept on a tight rein, with policymakers giving up a little slack each time the metaphoric dog growled too loudly. The collective constraints of restrictive trade union legislation and diluted employment rights appeared as the trophies of voluntarism and neo-liberal economic principles. In addition to this was an observed apathy towards the enforcement of legislation. This was highlighted in Chapter Three where it was shown that despite the prominence of positive statements on OHS, it was often the case that policymakers did not provide the means to honour stated intentions, through, for example, proactive enforcement-agency action that was independent from 'business-friendly' exigencies. In addition, it was argued that improvements in OHS practice and outcomes in the UK were impeded by the level, degree and range of employee involvement in OHS. While a full menu of employee involvement initiatives was on offer, most of them had the same bland flavour of managerial prerogative − a clearly potent and overwhelming ingredient in the contemporary employment relationship (and HRM). The limitations of employee involvement mechanisms underline the major changes that have occurred over the last two decades in the structure and organisation of the economy, in the labour market and in the position of organised labour. In turn, it was argued, these developments call for quite radical change to existing arrangements for OHS and employment rights in the UK, particularly in terms of the role and scope of safety representatives. For many commentators (for example, James and Walters 2002; Nichols 1997), the recalibration of power in the employment relationship is well overdue, but with managerial prerogative appearing as a cornerstone of HRM, such an exercise is likely to be well outside the range of the HRM machine.

In Chapter Four, the focus shifted to the relationship between work organisation and HRM policies and practices on work intensification at both an emotional and a physical labour level. Here we observed the effectiveness of HRM principles, policies and practices in maximising (short-term) employee contribution, although this often operated at the expense of employee health and well-being. In the case-study chapters on the international airline and call centre industries, firm linkages were established between policies covering work organisation (and the working environment), and employee health and safety outcomes. For airline cabin crews, the high incidence of various symptoms and illnesses were directly attributed by survey respondents to poor-quality cabin air, working patterns and work schedules. In addition to this, the international literature on call centre agent health and safety reported disturbingly high levels of MSDs and stress-related illnesses, all of which were intimately linked to high workloads, performance targets, electronic monitoring and rigorous working patterns. The role of management policy in employee health and safety was further implicated in 'organisational violence' (Bowie 2002), where, to a certain degree, management in some call centres (along with airlines and railway companies) create fertile conditions for customer violence, which in many cases they failed to address, either through the removal of the triggers, or by providing adequate training for employees. Given the strong linkage between the exposure to verbal abuse from customers and the higher incidence of stress-related symptoms, customer violence provided a further illustration of how management policy and practice can perpetuate and exacerbate unhealthy working conditions and illness amongst staff. In addition, the rising trend of customer violence in the interactive service industries directed our attention to the importance of incorporating the emotional aspects of labour into contemporary OHS debates.

The international call centre industry case study also delivered an interesting account of how HRM principles, policies and practices are mutating to fit with contemporary workplaces and the pressures within them. The literature illustrated how call centre managers have adapted a range of HR practices to meet the demands of customers and employees, while at the same time retaining a high degree of control over productivity and quality in the production process. Kinnie *et al.*'s (2000) study demonstrated how high-commitment policies may be used to balance conflicting pressures between the demands of the product market and the pressures of the labour market, allowing companies to satisfy the customer, improve business performance and maintain good relations with employees in a fast-paced and highly-controlled working environment. Standing well apart from the popular polarised views of the 'dark satanic mill' and the 'empowered knowledge worker', the authors suggested that what is emerging is an organisational form of tight technical and procedural controls that operate alongside high-commitment practices (Kinnie *et al.* 2000: 982). However, a more menacing account of HR practices was provided by Wallace *et al.*'s (2000) case studies, which demonstrated the exploitative potential of a range of sophisticated HR policies within a 'sacrificial HR strategy'. Common to both accounts was the facilitative role of HR practices in sustaining high productivity and obedience within work regimes that were generally depicted as 'intensive and stressful', even in 'model' call centres.

The industry case studies also provided evocative illustrations of the forces of political, economic and social pressures on OHS policy, practice and outcomes. Of particular interest was the variance in the level of legislative coverage in each of these industries, as well as the degree of scrutiny from enforcement agencies. For example, in the international airline industry, there appeared to be a relatively loose regulatory framework, most notably in terms of provisions for cabin crew health and safety. This combined with almost no enforcement activity – for example, no routine inspection of cabin air quality and hygiene standards. This was 'self-regulation' in its element. In comparison, the international call centre industry, particularly in First-World countries, was more highly regulated in terms of a greater number of applicable regulations and (possible) scrutiny from enforcement agencies. At the top of the regulations and monitoring tree was the international nuclear power industry, a position explained by its 'high-risk' classification – high risk in terms of both the potential for apocalyptic accidents, and the risk of losing a critical level of public support, which could severely limit future operations. A comparison of employee health outcomes between the three industries is not possible on the basis of the evidence we have presented, but we can speculate on the usefulness of stringent regulations and rigorous enforcement for 'good' OHS management. Even with some of the most extensive regulations and involvement by the enforcement agency, the UK nuclear power industry still managed to be criticised on no less than 143 counts of mismanagement and limited compliance. A puzzling finding indeed, suggesting that legislation and enforcement activity are not the only factors needed to resolve the OHS equation. Perhaps the missing variables can be located in 'safety culture'? Our assessment of 'safety culture' in Chapter Seven brought us back to how management and employees interact in OHS, for example, the importance of attitudes, behaviours, communication, clear responsibilities, involvement, consensus, and then it all started to sound like HRM again and the aroma of managerial prerogative began filtering through.

One important missing variable that has historical validity is meaningful employee participation – participation as opposed to 'involvement'. These two things are not the same (see, for example, Hyman and Mason 1995), and their differences relate to dimensions of both depth and power. A dictionary definition of 'participation' is 'to get actively involved in'. Team briefings, one of the most popular involvement techniques in the UK (Cully *et al.* 1999), could hardly be described on those terms. However, 'active participation' is all a bit of a waste of time if it is ignored. Specific rights backing participation along with the means (and will) to enforce and uphold these rights would go some way to ensure that employees' knowledge, experience and interests feature more conclusively in decisions over workplace health and safety management.

While tapping into and utilising employees' knowledge and experience offers considerable value in the management of workplace health and safety, knowledge of the job does not appear to be enough. As Vassie's (1998) case study showed, those involved in proactive health and safety management need specific and focused training on risk identification and resolution. Resourcing such training presents a significant hurdle for management, and it was noted that employers believed that training costs were the most inhibiting factors when planning improvements to workplace health and safety. While the UK government is presently reviewing trade union training grants that could be used for safety training, James and Walters (2002: 151) are quick to point out that not all safety representatives are members of trade unions. Indeed, judging by union density figures and the proportion of union-appointed safety representatives, non-union safety representatives appear as the majority, particularly if we consider the small-business sector. An all-encompassing funding package for all safety representatives is likely, therefore, to require substantial funding (perhaps one of the reasons why a half-measure approach is being considered), but wouldn't this be a better way to spend tax-payers' money as opposed to compensating (some) injured workers?

As alluded to earlier, a further contingency for more substantial improvements to OHS is likely to be the operation of well-resourced and empowered enforcement agencies covering all industrial sectors (that is, not just those classified as 'high risk'). Indeed, the value of enforcement agency activity is evident when we consider the catalyst effect of the NII in parts of the UK nuclear power industry. This would, of course, require significant support and funding from government as well as an ideological shift towards greater levels of intervention in employers' activities. The prospects for greater regulation of the employment relationship and OHS, as well as more rigorous enforcement activity, are, however, blackened by the UK government's obstinate position on the value of a voluntaristic, self-regulatory approach. As noted earlier, their position is that 'The [HSAW] Act provides a framework for good, effective regulation that has transformed Britain's workplaces' (Foreword in *Revitalising Health and Safety: Strategy Statement*, DETR/HSC, June 2000).

Not according to our account. However, there is one truth in the statement – workplaces have been transformed, but to something rather different than what the government purports. The government's vision for OHS is well-illustrated by the airline industry, which appears to be making the most of the situation. This is evidenced by the absence of provisions on cabin crew health and comfort and a range of minimalistic remedial measures for the areas where they have been caught out (for example DVTs). Of particular concern is the evidence that suggests airlines have been aware of oil leaks leading to organophosphate mists for at least a decade, paralleling the position of tobacco companies, who for decades buried the evidence of smoking-related cancers. We can only wonder about what other risks cabin crews and passengers are exposed to that we don't yet know about.

Based on the international airline industry case study, it would appear Neil Kinnock's judgement is as astute today as it was in the 1970s: 'History and reality, the past and the present, argue very forcibly against the idea that British industry is yet fit for self-regulation' (quoted in Hansard, vol. 875, May 1973). However, while it is clear that current regulatory frameworks, relatively limited employee rights and enforcement activities are primary explanatory variables in accounting for the abysmal OHS statistics in the UK (and elsewhere), the motivations and rationale of employers is not as clear-cut. In explaining their behaviour – typified by low levels of compliance to legislation, resistance to further regulatory and employee intervention in the employment relationship and the wide use of practices that create intensive and stressful work regimes – it is not enough to say that they do this 'because they can'. It could be the case that the conflicts and contradictions between the constant barrage of virtuous statements on OHS from policymakers and what we observe in practice are the result of a straightforward underestimation of the interaction between OHS and a range of people-management policies. This would help to explain why so many contradictions existed between the service, safety and quality goals in, for example, the international airline industry. Alternatively, a conspiracy theory might propose that OHS is knowingly exploited in pursuit of capital gain, meaning that in fact there is no conflict since the policies create the most appropriate circumstances that enable capital to achieve its goals. If we consider the bare-faced flouting of regulations in the UK call centre industry, and to some extent the international nuclear power industry, combined with the cynical utilisation of HRM principles, policies and practices, this position might be supported. A more balanced perspective might hold that the conflict and contradictions observed between publicly aspired-to goals and actual practice are more to do with the prevailing economic and competitive climates that restrict policymakers' options and limit employers' abilities to honour policy statements.

Whichever explanation fits best with any particular organisation, the fact remains that HRM principles, policies and practices along with notions of 'partnership' within a voluntaristic, self-regulatory framework are failing to provide the conditions that are required for responsible and high-quality health and safety management. By now, we are well versed on the main flaws and weaknesses leading to these inadequacies, as well as some of the possible remedial actions that could be taken. Based on the conclusions presented here, it is argued that the way forward in OHS is not about 'revitalising' health and safety, but more about re-engineering the framework within which OHS management operates. While clearly an anathema for some governments and employers, it is difficult to ignore the historical and more recent trends that depict managerial prerogative in the employment relationship as being antithetical to positive OHS outcomes. Based on the evidence presented, it appears that the combined forces of enhanced legislated employment rights, community action, empowered enforcement agencies free of 'dual-missions' and democratic forums for managing workplace health and safety, would be key components of any new framework – the enactment of which would not only challenge incumbent notions of profit-accumulation strategies, but also the philosophy and principles of the HRM concept. However, perhaps then some of the more humanistic working principles and practices adapted by HRM from its predecessors, 'organisational development' (OD) and the 'quality of working life' (QWL) movement, would be better placed to exert a more positive effect on workplace health and safety.

Notes

2 The regulatory politics of OHS

1 Department of Transport 1987, 1989; Department of Energy 1990.
2 The British model influenced the Industrial Safety, Health and Welfare Acts that were passed in Australian States: South Australia (1972), Tasmania (1977), Victoria (1981) and the 1983 Occupational Health and Safety Act in New South Wales (Gunningham 1985: 26).
3 Defining 'small' takes on different forms in different countries. A European recommendation defines 'small firms' as micro, small or medium-sized enterprises. In Britain, firms that employ fewer than 200 workers are defined as small, while in France, Germany and the USA, the figure is 500 (Odaka and Sawai 1999).
4 Subcontractors were implicated in the Hatfield rail crash in 2000.
5 While Valujet is a US registered company, the philosophy of self-regulation is equally prominent in the USA, and arguably led the way to UK deregulation of health and safety.
6 No deadline has been set for the completion of the installation. UK railways are only required to show that they are making progress towards the full installation of ETCS or its equivalent.
7 Investigators believed the crash was due to a fire sparked by oxygen generators in the jet's forward cargo hold. It was alleged the hazard could have been prevented had the subcontractor fitted a three-cent safety cap to the generators. The Acting US Attorney Guy Lewis described this as a clear case of 'putting corporate profits before safety'.
8 Compliance with regulations relating to age and disability do not come into force until 2 December 2006.

3 The social processes of OHS

1 Measured by working days lost due to labour disputes per 1,000 employees.
2 Up until July 2001, the maximum that a tribunal would force a vexatious employee to pay towards an employer's costs was £500. However, this limit has now been raised to £10,000, while the grounds for ordering such a payment have been broadened. The National Citizens' Advice Bureau has reported that some employees involved in forthcoming tribunals have received aggressive letters from their employers' solicitors warning them that they could be liable for thousands of pounds' worth of costs.
3 For the first time in the survey series, WERS (1998) examined employment relations in organisations with less than 25 employees.
4 The HSE main coverage is in factories, building sites, mines, farms, fairgrounds, quarries, railways, chemical plants, offshore and nuclear installations and hospitals. Local-authority enforcement officers cover retailing, some warehouses, most offices,

hotels and catering, sports, leisure, consumer services and places of worship – accounting for some 12,020,000 premises.

5 The 1977 Safety Representatives and Safety Committees (SRSC) Regulations entitled trade-union-appointed safety representatives to time off with pay to undertake training. This training used to be publicly funded through the Trade Union Training and Education Grant, but this was phased out by 1996.

6 The consultation exercise covered employers, including small and medium-sized enterprises (SMEs), and workers. Employers n = 194 responses; SMEs n = 134 responses; employees n = 860 responses.

4 Workplace factors in OHS

1 UNISON is a large public sector trade organisation.

2 Meta-analysis is a set of systematic techniques for quantitatively combining studies and applying statistical analyses to aggregated results.

3 All subjects were shift workers.

4 From data in the US Census over a 3-year period.

5 The train driver of the Southall train crash faced prosecution. The court case took place in June 1999 and found the driver not guilty. Great Western Trains were fined £1,500,000, but did not face manslaughter charges.

6 1998/99 figures from the HSC estimate that only 46 per cent of workplace injuries are reported under the Reporting Injuries, Diseases and Dangerous Occurrences Regulations (RIDDOR) 1995.

7 Acts of physical violence are now reportable under RIDDOR 1995.

8 No fatal injuries were recorded as a result of a physical attack on a worker (HSE, 2001a).

5 HRM and OHS in the international airline industry

1 The JAR operations manual details the full range of regulations covering aircraft and aircrew. These are too extensive to discuss in detail.

2 And the UK's CAA.

3 See, for example, Smith 1996, Nagda *et al.* 2000, Pukkala *et al.* 1995, Waters *et al.* 2000, Aspholm *et al.* 1999, Dorgan *et al.* 1999.

4 Following a two-year legal battle supported by the ITF and the French SNPNC union, the crews won their jobs back in April 1999 (*ITF News*, March 1998).

5 Nearly 2,000 called in sick on 11 July 1997 (BA figures).

6 The House of Lords Select Committee on Science and Technology, *Air Travel and Health Evidence*, 2000: 46, paragraphs 21 and 24.

7 The House of Lords Select Committee on Science and Technology, *Air Travel and Health Evidence*, 2000.

8 The House of Lords Select Committee on Science and Technology, *Air Travel and Health Evidence*, 2000: 46, paragraph 24.

9 The House of Lords Select Committee on Science and Technology, *Air Travel and Health Evidence*, 2000: 39, paragraph 2.

10 The House of Lords Select Committee on Science and Technology, *Air Travel and Health Evidence*, 2000: 34, paragraph 40.

11 The House of Lords Select Committee on Science and Technology, *Air Travel and Health Evidence*, 2000: 226, evidence from Diamond Scientific Ltd, supplier of air filters used on aircraft.

12 Airlines are required to report significant passenger illness to the port health authorities, for example if a large number of passengers suffered vomiting or diarrhoea during a flight or there are obvious signs of disease on a passenger, for example, chicken-pox sores. However, not all infectious diseases are so obvious.

13 Unit of measurement.
14 The hydraulic fluids and engine oils used in aircraft contain organophosphates.
15 Statement from the AFA, in The House of Lords Select Committee on Science and Technology, *Air Travel and Health Evidence*, 2000: 246.
16 The House of Lords Select Committee on Science and Technology, *Air Travel and Health Evidence*, 2000: 258.
17 The House of Lords Select Committee on Science and Technology, *Air Travel and Health Evidence*, 2000: 33, paragraph 36.
18 Farrol Kahn of the Aviation Health Institute (Oxford) is currently compiling a database of passenger reports of blood clots during/following air travel.
19 Professor Kevin Burnard is conducting a study of 2,000 passengers travelling from Britain to Australia to establish how many blood clots are recorded. Guy's and St Thomas's Hospitals, London.
20 In a recent UK legal suit, a businessman's claim was upheld against JMC travel for suffering significant discomfort during a flight from London to Calgary. The judge said there should be a minimum distance of 34 inches between seats on long-haul flights (*Guardian*, 17 April 2002).
21 *The Government's Response to the Report of The House of Lords Select Committee on Science and Technology: Air Travel and Health*, 2001: 7.
22 Ninety eight per cent reported that they 'often' or 'sometimes' worked mixed shifts entailing both very early and very late starts.
23 The three airlines in the study said that cabin crew receive at least five days of training per year.

6 HRM and OHS in the international call centre industry

1 Taylorism relates to low-skill, low-control tasks that are subject to strict monitoring. See Taylor (1947).
2 Display-screen equipment (DSE) includes visual-display units, keyboards, mice and laptop computers.
3 See Foucault (1977) for a full discussion.
4 The findings are based on a sample of 7,268 employers in establishments with a minimum of fifty employees based in eighteen European countries (the fifteen EU member states plus Hungary, Poland and the Czech Republic).
5 The HSE was due to produce a Contract Research Report (number 9678) based on the findings of the 2001 HSE/HELA report on call centres. This was not available at the time of writing.
6 The HSC document states that the equivalent statistics for notices issued by local-authority inspectors are not available (HSC 2000b: 9).

7 HRM and OHS: safety culture in the international nuclear power industry

1 Focus groups (four groups of between six and eight people) and telephone interviews (n = 1006) were conducted between September 1999 and March 2000 across ten UK locations that varied in proximity to nuclear installations. The sample was considered representative of gender and age.
2 Level 4 denotes an accident without significant off-site risks (IAEA International Nuclear Event Scale).
3 The Ionising Radiation Regulations 1999.
4 The plant was originally constructed to reprocess Dounreay Fast Reactor fuel but was refurbished to carry out the dissolution and reprocessing of mixed (uranium and plutonium) oxide fuel (MOX) for the Prototype Fast Reactor (HSE 1998b: 107).
5 At the end of June 2001, 60 of the 143 recommendations had been fulfilled or 'closed out'.

Bibliography

Abel, R. (1985) 'Risk as an arena of struggle', *Michigan Law Review*, February: 77–812.

Ackers, P. (2001) 'Employment ethics', in T. Redman and A. Wilkinson (eds) *Contemporary Human Resource Management: Texts and Cases*, Harlow: FT/Prentice Hall, Pearson Education.

Advisory Committee of the Safety of Nuclear Installations (ACSNI) (1993) Study Group on Human Factors, *Third Report: Organising for Safety*, London: HMSO.

Agho, O. A., Mueller, C. W. and Price, J. L. (1992) 'Discriminant validity of measures of job satisfaction, positive affectivity and negative affectivity', *Journal of Occupational and Organisational Psychology* 65: 185–96.

Agius (2001) *Health, Environment and Work*, http://www.agius.com.

Air Transport World, May 1997.

Åkerstedt, T. (1994) 'Work injuries and time of day', Proceedings of the Work Hours, Sleepiness and Accidents Symposium, Stockholm, Sweden, 8–10 September 1994: 106.

Albrow, M. (1994) 'Accounting for organisational feeling', in L. Ray and M. Reed (eds) *New Weberian Perspectives on Work, Organisation and Society*, London: Routledge.

Albrow, M. (1997) *Do Organisations Have Feelings?*, London: Routledge.

Arrowsmith, J. and Sisson, K. (2001) 'Managing working time', in S. Bach and K. Sisson (eds) *Personnel Management: A Comprehensive Guide to Theory and Practice*, Oxford: Blackwell.

Ashforth, B. E. and Humphrey, R. H. (1993) 'Emotional labour in service roles: the influence of identity', *Academy Management Review* 18: 88–115.

Ashkan, K., Nasim, A., Dennis, M. J. S. and Sayers, R. D. (1998) 'Acute arterial thrombosis after a long-haul flight', *Journal Royal Society of Medicine* 91: 324.

Aspholm, R., Lindbohm, M. L., Paakkulainen, H., Taksinen, H., Nurminen, T. and Tiitinen, A. (1999) 'Spontaneous abortion among Finnish flight attendants', *Journal of Occupational and Environmental Medicine* 41(6): 486–91.

Association of Flight Attendants (AFA) (1997) 'Air Safety', http://www.flightattendant-afa.inter.net/

Bach, S. (1994) 'The working environment', in K. Sisson (ed.) *Personnel Management*, 2nd edition, London: Blackwell.

Bagshaw, M. (1998) Untitled Presentation, Medical University of South Carolina, Environmental Hazards Assessment Conference, 5–7 February 1998: 12–17.

Bagshaw, M., Irvine, D. and Davies, D. M. (1996) 'Exposure to cosmic radiation of British Airways flying crew on ultralonghaul routes', *Occupational and Environmental Medicine* 53(7): 495–8.

Bain, P. (1997) 'Human resource malpractice: the deregulation of health and safety at work in the USA and Britain', *Industrial Relations Journal* 28(3): 176–89.

Bain, P. (2001) 'Some sectoral and locational factors in the development of call centres in the USA and the Netherlands', Occasional Paper 11, Department of Human Resource Management, University of Strathclyde.

Bain, P. M. and Baldry, C. J. (1995) 'Sickness and control in the office – the Sick Building Syndrome', *New Technology, Work and Employment* 10(1): 19–31.

Baldry, C. and Bain, P. (1994) 'Trade unions and Sick Building Syndrome', April, Trade Union Report No. 2, University of Strathclyde.

Balouet, J. C. (2000) Evidence to the House of Lords Select Committee on Science and Technology: Air Travel and Health, 245.

Band, P. R., Le, N. D., Fang, R., Deschamps, M., Coldman, A. J., Gallagher, R. P. and Moody, J. (1996) 'Cohort study of Air Canada pilots: mortality, cancer incidence and leukaemia', *American Journal of Epidemiology* 143: 137–43.

Barish, R. J. (1999) 'Human Resource Manager responsibilities with respect to business frequent flyer radiation exposure', unpublished paper.

Barker, G. (1998) 'Factories of the future', The *Age*, 24 February 1998.

Barton, J. and Folkard, S. (1993) 'Advancing versus delaying shift systems,' *Ergonomics* 36(1–3): 59–64.

BBC Online Health, 29 June 1999 'Radiation hazards for frequent flyers', http://www.bbc.co.uk

Beaumont, P. B., Coyle, J. R., Leopold, J. W. and Schuller, T. E. (1982) *The Determinants of Effective Joint Health and Safety Committees*, University of Glasgow, Centre for Research into Industrial Democracy and Participation.

Beck, M. and Woolfson, C. (1999) 'Safety culture – a concept too many?', *The Health and Safety Practitioner* 16(1): 14–16.

Belt, V. (2000) 'Call girls: women, work and the telephone', paper presented to the 18th Annual International Labour Process Conference, 25–7 April, University of Strathclyde.

Belt, V., Richardson, R. and Webster, J. (2000) 'Women's work in the information economy: the case of telephone call centres', *Information, Communication and Society* 3(3): 366–85.

Blair, A., Hartge, P., Stewart, P. A., McAdams, M. and Lubin, J. (1998) 'Mortality and cancer incidence of aircraft workers exposed to trichloroethylene and other organic solvents and chemicals', *Occupational and Environmental Medicine* 55(93): 161–71.

Blossfeld, H. P. and Hakim, C. (eds) (1997) *Between Equalization and Marginalization: Women Working Part-time in Europe and the United States of America*, Oxford: Oxford University Press.

Blyton, P., Lucio, M. M., McGurk, J. and Turnbull, P. (1998) *Contesting Globalization: Airline Restructuring, Labour Flexibility and Trade Union Strategies*, Cardiff University and the International Transport Workers' Federation.

Blyton, P. and Turnbull, P. (1995) *Reassessing Human Resource Management*, London: Sage.

Blyton, P. and Turnbull, P. (1998) *The Dynamics of Employee Relations*, 2nd edition, London: Macmillan.

Bohle, P., Quinlan, M. and Mayhew, C. (2001) 'The health effects of unemployment and job insecurity: an old debate and new evidence', Industrial Relations Research Centre, University of New South Wales, Australia.

Bolton, S. C. (2000) 'Who cares? Offering emotion work as a "gift" in the nursing labour process', *Journal for Advanced Nursing* 32(3): 580–6.

Bolton, S. C. (2001) 'Changing faces: nurses as emotional jugglers', *Sociology of Health and Illness* 23(1): 85–100.

Bolton, S. and Boyd, C. (2003) 'Trolley dolly or skilled emotion manager? Moving on from Hochschild's managed heart', *Work, Employment and Society*, 17(2).

Bowie, V. (2002) 'Defining violence at work: a new typology', in M. Gill, B. Fisher and V. Bowie (eds) *Violence at Work: Causes, Patterns and Prevention*, Devon: Willian Publishing.

Boyd, C. (1996) 'A sick environment in the sky? Aircraft and sick building syndrome', unpublished BA Dissertation, University of Strathclyde.

Boyd, C. (2001) 'Managing our most important asset: the rhetoric and reality of HRM in the airline industry', Ph.D. Dissertation, University of Strathclyde.

Boyd, C. (2002) 'Customer violence and employee health and safety', *Work, Employment and Society* 16(1): 151–69.

Boyd, C. and Bain, P. (1998) ' "Once I get you up there where the air is rarified": health, safety and the working conditions of airline cabin crews', *New Technology, Work and Employment* 13(1): 16–28.

Boyd, C. and Bain, P. (1999) 'The new working life of airline cabin crews', *Proceedings of the International Conference of the Institute of Working Life*, 11–13 January, Stockholm.

Braunholtz, S., McDonald, I. and Elgood, J. (1998) *Health and Safety in the Workplace*, London: MORI/HSE, HSE Books.

Braverman, H. (1974) *Labour and Monopoly Capital: The Degradation of Work in the Twentieth Century*, New York: Monthly Review Press.

British Airways (2000) *Annual Report*.

Brooke, P. B. (1986) 'Beyond the Steers and Rhodes model of employee attendance', *Academy of Management Review* 11: 345–61.

Brundrett, G. (2001) 'Comfort and health in commercial aircraft: a literature review', *Journal of the Royal Society for the Promotion of Health* 121(1): 29–37.

Buell, P. and Breslow, L. (1960) 'Mortality from coronary disease in California men who work long hours', *Journal of Chronic Diseases* 11: 615–26.

Building Research Establishment Ltd (BRE) and Department for Transport, Local Government and the Regions (DTLR) (2001) 'Health in Aircraft Cabins', Stage 2 Report, September 2001, http://www.aviation.dtlr.gov.uk/healthcab/cabinhealth

Burgess, J. (1992) 'Further evidence on small business employment and industrial relations', *Labour Economics and Productivity* 4(2): 85–102.

Burnard, K. (1999) (quoted in the *Sunday Times*, 29 July 1999), Study of 2,000 passengers on long flights suffering clots during travel between Britain and Australia, Guy's and St Thomas's Hospitals, London.

Burrows, N. and Woolfson, C. (2000) 'Regulating business and the business of regulation; the encouragement of business friendly assumptions in regulatory agencies', in L. MacGregor, T. Prosser and C. Villiers (eds) *Regulation and Markets Beyond 2000*, Aldershot: Ashgate, 319–40.

Call Centre News (1998) *International Call Centre Statistics*.

Callaghan, G. and Thompson, P. (2001) 'Edwards revisited: technical control and call centres', *Economic and Industrial Democracy* 22: 13–37.

Cappelli, P. (ed.) (1995) *Airline Labor Relations in the Global Era*, New York: ILR Press/Cornell.

Carson, W. G. (1985) 'Hostages to history: some aspects of the occupational health and safety debate in historical perspective', in W. B. Creighton and N. Gunningham (eds) *The Industrial Relations of Occupational Health and Safety*, Sydney: Croom Helm.

Carson, W. G. (1989) 'Occupational health and safety: a political perspective', *Labour and Industry* 2(2).

Carson, W. G. and Henenberg, C. (1988) 'The political economy of legislative change: making sense of Victoria's new health and safety legislation', *Law in Context* 6(2).

Cauldwell, J. A. Jnr (1997) 'Fatigue in the aviation environment: an overview of the causes and effects as well as the recommended countermeasures', *Aviation, Space and Environmental Medicine* 68(10): 932–8.

Channel Four, *Dispatches*, 6 February 1998.

Channel Four, *4x4*, 23 January 2001.

Chappell, D. and Di Martino, V. (1998) *Violence At Work*, Geneva: International Labour Organisation (ILO).

Chartered Institute of Personnel and Development (CIPD) (1995) *Occupational Health and Organisational Effectiveness*, London: CIPD.

Cheyne, A., Cox, S., Oliver, A. and Tomas, J. M. (1998) 'Modelling safety climate in the prediction of levels of safety activity', *Work and Stress* 12: 255–76.

Civil Aviation Authority (CAA) (1999) Notes from an interview with Captain Mike Vivan, Flight Safety.

Clark, S. P. and Borders, L. D. (1984) 'Approaches to sexual and sexually harassing callers', *Crisis Intervention* 1: 14–24.

Clark, S. P., Borders, L. D. and Knudson, M. L. (1986) 'Survey of telephone counsellors' responses to sexual and sexually abusive callers', *American Mental Health Counselors Association Journal* 2: 73–9.

Clarke, S. G. (1999) 'Perceptions of organisational safety: implications for the development of safety culture', *Journal of Organisational Behaviour* 20: 185–98.

Clarke, S. G. (2000) 'Safety culture: under-specified and overrated?', *International Journal of Management Reviews* 2(1): 65–90.

Cohen, A. (1997) 'Factors in successful safety programs', *Journal of Safety Research* 9(4): 168–78.

Collinson, D. L. (1992) *Managing the Shopfloor: Subjectivity, Masculinity and Workplace Culture*, Berlin: Walter de Gruyter.

Confederation of British Industry (CBI) (1990) *Developing a Safety Culture*, London: CBI.

Cooper, C. L. (1985) 'The stress of work: an overview', *Aviation, Space and Environmental Medicine* 56(70): 627–32.

Cottrell, J., Lebovitz, B., Fennel, R. and Kohn, G. (1995) 'Inflight arterial saturation continuous monitoring by pulse oximetry', *Aviation, Space and Environmental Medicine*, February: 126–30.

Cox, S. and Flin, R. (1998) 'Safety culture: philosopher's stone or man of straw?', *Work and Stress* 12(3): 189–201.

Cully, M., O'Reilly, A., Millward, N., Woodland, S., Dix, G. and Bryson, A. (1999) *Britain at Work: As Depicted by the 1998 Workplace Employee Relations Survey*, London: Routledge.

Cutler, T. and James, P. (1996) 'Does safety pay? A Critical Account of the Health and Safety', *Work, Employment and Society* 10(40).

Danford, A. (1998) 'Work organisation inside Japanese firms in South Wales: a break from Taylorism?', in P. Thompson and C. Warhurst (eds) *Workplaces of the Future*, London: Routledge.

Daniel, W. E., Vaughan, R. L. and Millies, B. A. (1990) 'Pregnancy outcomes among female flight attendants', *Aviation, Space and Environmental Medicine*, September: 840–4.

Datamonitor (1999) *Opportunities in US and Canadian Call Centre Markets: The Definitive Vertical Analysis*, New York: Datamonitor.

Datamonitor (2001) *Call Centres in Europe*, London: Datamonitor.

Dawood, R. M. (2000) Evidence to the House of Lords Select Committee on Science and Technology: Air Travel and Health, 221.

Deery, S., Iverson, R. and Walsh, J. (2000) 'Work relationships in telephone call centres: understanding emotional exhaustion and employee withdrawal', paper to the International Industrial Relations Conference, Manchester, February.

Delbridge, R. and Turnbull, P. (1992) 'Human resource maximization: the management of labour under just-in-time manufacturing systems', in P. Blyton and P. Turnbull (eds) *Reassessing Human Resource Management*, London: Sage Publications.

Department of Energy (1990) *Public Enquiry into the Piper Alpha Disaster*, London: HMSO.

Department of the Environment (1997) *Review of Past Nuclear Accidents: Source Terms and Recorded Gamma-ray Spectra*, London: DOE.

Department of the Environment, Transport and the Regions (DETR) and the Health and Safety Commission (HSC) (2000) *Revitalising Health and Safety: Strategy Statement*, June, London: DETR.

Department of Health (1999) *The Healthy Workplace Initiative*, April, London: HMSO.

Department of Trade and Industry (DTI) (1994) *Thinking About Regulation: A Guide to Good Regulation*, London: DTI.

Department of Trade and Industry (1998) *Small and Medium-sized Enterprise Statistics for the UK*, London: DTI.

Department of Trade and Industry (2001) *Implementation of the Working Time Regulations*, Employment Relations Research Series No. 11, London: DTI.

Department of Transport (1987) *Formal Investigation of the* Herald of Free Enterprise: *Report of Court 8074*, London: HMSO.

Department of Transport (1989) *Formal Investigation of the Clapham Junction Rail Accident*, London: HMSO.

Department for Transport, Local Government and the Regions (2001) *Disruptive Behaviour on Board UK Aircraft: Analysis of Incident Reports April 2000–March 2001*.

Dolley, S. (1999) 'Japan's Nuclear Criticality Accident: N-base Tokai Accident Reports', http://www.n-base.org.uk

Donnelly, M. (1997) *An investigation of gastrointestinal illness among airline passengers*, Public Health Laboratory Services.

Donnini, G., Van Hiep Nguyen and Haghighat, F. (1990) 'Ventilation control and building dynamics by CO_2 measurement', Proceedings of the 5th International Conference on Indoor Air Quality and Climate, Toronto, Canada.

Dorgan, J. F., Brock, J. W., Rothman, N., Needham, L. L., Miller, R., Stephenson, H. E. Jnr, Schussler, N. and Taylor, P. R. (1999) 'Serum organochlorine pesticides and PCBs and breast cancer risk', *Cancer Causes and Control* 10(1): 1–11.

Dorman, P. (1988) 'A negotiable "workers-rights" model of occupational safety policy', *International Review of Applied Economics* 2(2): 170–88.

Dorman, P. (1996) *Markets and Mortality*, Cambridge: Cambridge University Press.

Driver, C. R., Valway, S. E., Morgan, W. M., Onorato, I. M. and Castro, K. G. (1994) 'Transmission of M. tuberculosis associated with air travel', *JAMA* 272: 1031–5.

Economist Intelligence Unit (1999) 'The world in figures: countries', in D. Fishburn (ed.) *The World in 1999*, London: The Economist Publications.

Economist National Safety Council (1999).

Erickson, R. J. and Wharton, A. S. (1997) 'Inauthenticity and depression: assessing the consequences of interactive service work', *Work and Occupations* 24(2): 188–213.

European Commission (1997) *Air Transport Report prepared by Cranfield University*, London: Kogan Page.

European Foundation for the Improvement of Living and Working Conditions (EFILWC) (1997) *Preventing Absenteeism in the Workplace*, Dublin: European Foundation.

European Foundation for the Improvement of Living and Working Conditions (EFILWC) (1998) *EPOC Study*, EF/97/46/EN.

European Industrial Relations Review (2000a) 'Call Centres in Europe: Part One', September (320): 13–20.

European Industrial Relations Review (2000b) 'Call Centres in Europe: Part Two', September (321): 13–19.

European Industrial Relations Review (2001) 'Social Policy State of Play', January (324): 13–22.

Eurostat (1995) *Europe in Figures*, fourth edition.

Fairbrother, P. (1996) 'Organise and survive: unions and health and safety – a case study of an engineering unionised workforce', *Employee Relations* 18(2).

Fanger, O. (2001) 'Human requirements in future air-conditioned environments', *International Journal of Refrigeration* 24: 148–53.

Federal Aviation Administration (FAA) (1980) Advisory Circular AC: 120–38.

Felstead, A. and Jewson, N. (1999) *Global Trends in Flexible Labour*, London: Macmillan.

Fernie, S. (1998) 'Call centres – the workplace of the future of the sweatshops of the past in a new disguise', *Centrepiece*, London: Centre for Economic Performance, London School of Economics.

Fernie, S. and Metcalf, D. (1998) *(Not) Hanging on the Telephone: Payment Systems in the New Sweatshops*, London: Centre for Economic Performance, London School of Economics.

Financial Times (1999) 'Damages after nuclear accident', 30 December: 1.

Financial Times (2000a) 'Japan's quality control slips as corporate ethos changes', 28 July: 16.

Financial Times (2000b) 'Six arrests over Japan N-plant', 12 October: 11.

Financial Times Information Limited (2002) 'Budget airlines: too much of a good thing', 29 August, London: *Financial Times*.

Fine, G. A. (1998) 'Letting off steam? Redefining a restaurant work environment', in M. Jones, M. Moore and R. Snyder (eds) *Inside Organisations: Understanding the Human Dimension*, Newbury Park: Sage Publications.

Fineman, S. (1993) *Emotions in Organisations*, London: Sage.

Flight Attendants and Related Services (NZ) Association (FARSA) (2002) *Planetalk*, Issue 89, April.

Flournoy, A. C. (1990) 'Selected legal issues related to sound and vibration in pregnancy', in *Seminars in Perinatology* 14: 334–9.

Flynn, C. and Thompson, T. (1990) 'Effects of acute increases in altitude on mental status', *Psychometrics* 31(2).

Foucault, M. (1977) *Discipline and Punish: The Birth of a Prison*, London: Allen Lane.

Friedberg, W., Copeland, K., Duke, F. E., O'Brien, K. 3rd and Darden, E. B. (2000) 'Radiation exposure during air travel', *Radiation Protection Dosimetry* 48: 21–5.

Frisch, W. and Gros, G. (2001) 'Improving the safety of future nuclear fission power plants', *Fusion Engineering and Design* 56–7: 83–93.

Goldhagen, P. (2000) 'Overview of aircraft radiation exposure and recent ER-2 measurements', *Health Physics* 74: 526–44.

Goodhead, D. (1999) Cosmic Radiation Seminar, London, Aviation Health Institute.

Goodwin, T. (1996) 'Cabin crew maternity policy, the health and safety issues', www.aeronet.co.uk

Goss, D. and Adam-Smith, D. (2001) 'Pragmatism and compliance: employer responses to the working time regulations', *Industrial Relations Journal* 32(3): 195–208.

The Government's Response to the Report of The House of Lords Select Committee on Science and Technology: Air Travel and Health, February 2001, London: The Stationery Office, http://www.ukstate.com

Guardian, 26 July 1997.

Guardian, 5 August 1997.

Guardian (1998) 'Oneworld, five airlines: marriages abound among jet set', 22 September: 25.

Guardian (2000) '78m payout for radiation', 5 September: 17.

Guardian, 25 October 2000.

Guardian, 22 November 2000.

Guardian, 11 November 2001.

Guardian (2002) 'Strike threat as BA cuts 5800 jobs', 14 February: 2.

Guardian (2002) 'Qantas to start low-cost airline', 3 April: 26.

Guardian (2002) 'Airlines in hot seat on legroom ruling', 17 April: 7.

Guardian (2002) 'Easyjet lines up merger with Go: shake-up of budget airlines could mean higher fares', 4 May: 2.

Guardian (2002) 'US airline asks for \$1bn bailout', 18 May: 27.

Guardian (2002) 'Travel chaos: summer of discontent', 23 June: 8.

Gunderstrup, M. and Storm, H. (1999) 'Radiation induced leukaemia and other cancers in commercial jet cockpit crew: a population-based cohort study', *Lancet* 354: 2020–31.

Gunningham, N. (1985) 'Workplace safety and the law', in W. B. Creighton and N. Gunningham (eds) *The Industrial Relations of Occupational Health and Safety*, Sydney: Allen and Unwin.

Hackett, R. D. (1989) 'Work attitudes and employee absenteeism: a synthesis of the literature', *Journal of Applied Psychology* 66: 574–80.

Hansard, vol. 875, 21 May 1973.

Härikainen-Sörri, A. (1988) 'Occupational noise during pregnancy: a case control study', *Occupational and Environmental Health* 60: 279–83.

Härma, M., Suvanto, S. and Partinen, M. (1994) 'The effect of four-day round trip flights over 10 time zones on the sleep–wakefulness patterns of airline flight attendants', *Ergonomics* 37(9): 1461–78.

Harris, A. (1999) 'Violence in the workplace: what is the problem?', in *Final Report of Conference on Violence to Staff on Railways*, 8 July, RIAC/HSC.

Harris, W. and Mackie, R. R. (1972) *A Study of the Relationships Among Fatigue, Hours of Service and Safety of Operations of Truck and Bus Drivers*, Report 1727-2, Goleta, California: Human Factors Research Inc.

Hart, T. M. (1993) 'Human resource management: time to exorcise the militant tendency', *Employee Relations* 15(3): 29–36.

Harvey, J., Erdos, G., Bolam, H., Cox, M. A. A., Kennedy, J. N. P. and Gregory, D. T. (2002) 'An analysis of safety culture attitudes in a highly regulated environment', *Work and Stress* 16(1): 18–36.

Harvey, J., Bolam, H., Gregory, D. and Erdos, G. (2001) 'The effectiveness of training to change safety culture and attitudes within a highly regulated environment', *Personnel Review* 30(6): 615–36.

Health and Safety Bulletin (1997), 262, October, London: Industrial Relations Review.

Health and Safety Commission (HSC) (1993) *Third Report: Organising for Safety*, ACSNI Study Group on Human Factors, London: HMSO.

Health and Safety Commission (HSC) (2000a) *Health and Safety Statistics 1999/2000*, London: HSC.

Health and Safety Commission (HSC) (2000b) *Second Report to the EC on Practical Implementation of the Display Screen Equipment Directive*, London: HSC/00/P60.

Health and Safety Commission (HSC) (2001) *Health and Safety Statistics*, 24 October, http://www.hse.gov.uk

Health and Safety Executive (HSE) (1989) *Human Factors in Industrial Safety*, London: HMSO.

Health and Safety Executive (HSE) (1991) *Successful Health and Safety Management*, London: HMSO.

Health and Safety Executive (HSE) (1995) *Stress at Work: A Guide for Employers*, London: HSE Books.

Health and Safety Executive (HSE) (1997) *The Costs of Accidents at Work*, London: HSE Books.

Health and Safety Executive (HSE) (1998a) 'Display screen equipment health problems: user-based assessments of DSE health risks', HSE Contract Research Report 198/1998, London: HSE Books.

Health and Safety Executive (HSE) (1998b) *Safety Audit of Dounreay*, London: HSE Books.

Health and Safety Executive (HSE) (1999a) *Reducing Risks, Protecting People*, Discussion Document, May, London: HSE Books.

Health and Safety Executive (HSE) (1999b) *Violence at Work: Findings from the British Crime Survey*, London: HSE Books.

Health and Safety Executive (HSE) (2000a) *Health and Safety Offences and Penalties: A Report by the Health and Safety Executive*, London: HSE Books.

Health and Safety Executive (HSE) (2000b) *Regulating Higher Hazards: Exploring the Issues*, London: HSE Books.

Health and Safety Executive (HSE) (2000c) 'Statement of nuclear accidents at nuclear installations', Press Release, 4 October, London: HSE Books.

Health and Safety Executive (HSE) (2000d) *Regulating Higher Hazards: Exploring the Issues*, London: HSE Books.

Health and Safety Executive (HSE) (2000e) *Statement of nuclear accidents at nuclear installations*, Press Release 4 October.

Health and Safety Executive (HSE) (2001a) *Health and Safety Statistics*, London: HSE Books.

Health and Safety Executive (HSE) (2001b) *The Joint Inquiry into Train Protection Systems*, London: HSE Books.

Health and Safety Executive (HSE) (2001c) *Reducing Risks, Protecting People*, revised discussion document, first published in May 1999, London: HSE Books.

Health and Safety Executive (HSE) (2002) 'HSE and SEPA publish final report on Dounreay safety audit', Press Release, 22 January, London: HSE Books.

Health and Safety Executive/Health and Safety Executive Local Authority Enforcement Liaison Committee (HSE/HELA) (1999) *Initial Advice Regarding Centre Working Practices*, Report 94/1, November.

Health and Safety Executive/Health and Safety Executive Local Authority Enforcement Liaison Committee (HSE/HELA) (2001) *Advice Regarding Centre Working Practices*, Report 94/1 (rev.), December.

Health and Safety in Local Authority Enforced Sectors (HELA) (2000) *Annual Report 2000*, London: HSE Books.

Health and Safety in Local Authority Enforced Sectors (HELA) (2001) *Annual Report 2001*, London: HSE Books.

Herald (1995) 3 January.

Herald (1997) 7 June.

Hill, S. (1991) 'Why quality circles failed but total quality management might succeed', *British Journal of Industrial Relations* 29(4): 541–68.

Hirsch, B. T. and Berger, M. C. (1984) 'Union membership determination and industry characteristics', *Southern Economic Journal* 50.

Hochschild, A. R. (1979) 'Emotion work, feeling rules, and social structure', *American Journal of Sociology* 85: 551–75.

Hochschild, A. R. (1983) *The Managed Heart: Commercialization of Human Feeling*, Berkeley, California: University of California Press.

Hofstede, G. (1991) *Culture and Organizations: Software of the Mind*, New York: Harper-Collins.

Höpfl, H. G. (1994) 'Safety culture, corporate culture: organizational transformation and the commitment to safety', *Disaster Prevention and Management* 3: 49–58.

House of Lords Select Committee on Science and Technology (2000): *Air Travel and Health, Session 1999–2000, 5th Report*, 15 November 2000, London: The Stationery Office, http://www.ukso.co.uk

Howard, S. (1998) International Transport Workers' Federation, Press Release.

Hunter, L. and MacInnes, J. (1992) 'Employers and labour flexibility: the evidence from the case studies', *Employment Gazette*, June, 307–15.

Huws, U. and O'Regan, S. (2001) *E-Work in Europe: The EMERGENCE 18-country Employer Survey*, Institute for Employment Studies, Report No. 380.

Huws, U., Jagger, N. and O'Regan, S. (1999) *Teleworking and Globalisation*, Institute for Employment Studies, Report No. 358.

Hyman, J. and Mason, B. (1995) *Managing Employee Involvement and Participation*, London: Sage.

Hyman, R. (1995) 'The historical evolution of British industrial relations', in P. Edwards (ed.) *Industrial Relations: Theory and Practice in Britain*, Oxford: Blackwell.

Ichniowski, C., Kochan, T., Levine, D., Olsen, C. and Strauss, G. (1996) 'What works at work: overview and assessment', *Industrial Relations* 35(3): 299–333.

Income Data Services Ltd (1998) *Pay and Conditions in Call Centres*, London: IDS.

Income Data Services Ltd (2000) *Pay and Conditions in Call Centres*, London: IDS.

Independent (1997) 26 July.

IndiaOneStop.com (2001) 'Macro-economic overview: industrial production', http://www.indiaonestop.com

Industrial Relations Services (IRS) (1997) 677, April 1997.

Industrial Relations Services (IRS) (1999) 700, April 1999.

Institute for Employment Studies (1997) *Evaluation of the Display Screen Equipment Regulations 1992*, HSE Contract Research Report 130/1997, London: HSE Books.

Institute for Environment and Health (2001) *A Consultation on the Possible Effects on Health, Comfort and Safety of Aircraft Cabin Environments*, January, London.

Institute for Occupational Medicine (1999) 'Epidemiological and ergonomic study of occupational factors associated with symptoms of upper limb disorders in keyboard operators', Institute for Occupational Medicine, Edinburgh, October.

International Atomic Energy Agency (IAEA) (1996) *One Decade After Chernobyl: The Basis for Decisions*, Vienna: IAEA.

International Atomic Energy Agency (IAEA) (1998) General Conference on Measures to

Strengthen International Cooperation in Nuclear, Radiation and Waste Safety, 12 August, Vienna: IAEA.

International Atomic Energy Agency (IAEA) (1999) *Report on the Preliminary Fact Finding Mission Following the Accident at the Nuclear Fuel Processing Facility in Tokaimura, Japan*, Vienna: IAEA.

International Atomic Energy Agency (IAEA) (2001a) 'Nuclear Power Statistics', Press Release, 3 May, www.iaea.org

International Atomic Energy Agency (IAEA) (2001b) '15 Years After Chernobyl, Nuclear Power Plant Safety Improved Worldwide, But Regional Strains on Health, Economy and Environment Remain', Press Release, 25 April, www.iaea.org

International Transport Workers' Federation (ITF) (1995) *Cabin Crew: Safety Professionals*, London: ITF publications.

International Transport Workers' Federation (ITF) (1997) Media Release, 8 July.

International Transport Workers' Federation (ITF) (1998a) *Open Skies, Global Alliances and Labour Relations in the 21st Century*, London: ITF Publications.

International Transport Workers' Federation (ITF) (1998b) *Asian Crisis Threatens Aviation Safety – Staff Under Pressure*, September, London: ITF Publications.

Irvine, D. and Davis, D. M. (1999) 'British Airways flightdeck mortality study, 1950–1992', *Aviation, Space and Environmental Medicine* 71: 647–8.

ITF News 1 (1997) London: International Transport Workers' Federation.

ITF News 2 (1998) London: International Transport Workers' Federation.

ITF News 3 (1998) London: International Transport Workers' Federation.

ITF News 4 (1998) London: International Transport Workers' Federation.

ITF News 6 (1998) London: International Transport Workers' Federation.

ITF News 1 (1999) London: International Transport Workers' Federation.

ITF News 4 (1999) London: International Transport Workers' Federation.

James, P. (1992) 'Reforming British health and safety law: a framework for discussion', *Industrial Law Journal* 21(2).

James, P. and Walters, D. (1997) 'Non-union rights of involvement: the case of health and safety at work', *Industrial Law Journal* 26(1): 35–50.

James, P. and Walters, D. (2002) 'Worker representation in health and safety: options for regulatory reform', *Industrial Relations Journal* 33(2): 141–56.

Johnstone, R., Mayhew, C. and Quinlan, M. (2001) 'Outsourcing Risk? The Regulations of Occupational Health and Safety Where Subcontractors are Employed', Industrial Relations Research Centre, University of New South Wales, Australia.

Jones, A. P. (1999) 'Indoor air quality and health', *Atmospheric Environment* 33: 4535–64.

Kahn, F. (1999) Telephone conversation with the author, Oxford: The Aviation Health Institute.

Kahn, F. (2000) Evidence to the House of Lords Select Committee on Science and Technology: Air Travel and Health, 9 May.

Karasek, R. A. (1979) 'Job demands, job decision latitudes, and mental strain: implications for job redesign', *Administrative Science Quarterly* 24(2): 285–308.

Karasek, R. and Theorell, T. (1990) *Healthy Work: Stress, Productivity and the Reconstruction of Working Life*, New York: Basic Books.

Karr, A. R. (1987) 'GM agrees to pay fine of $500,000 to OSHA Charges', *The Wall Street Journal*, 6 October.

Karube, H., Aizawa, Y., Nakamura, K., Maeda, A., Hashimoto, K. and Takata, T. (1995) 'Oil mist exposure in industrial health – a review', *Sangyo Eiseigaku Zasshi* 37(2): 113–22.

Katz, J. E. (1994) 'Empirical and theoretical dimensions of obscene phone calls to women in the United States', *Human Communication Research* 21: 155–82.

Keenoy, T. (1997) 'HRMism and the languages of re-presentation', *Journal of Management Studies* 34(5).

Keenoy, T. and Anthony, P. (1992) 'HRM: metaphor, meaning and morality', in P. Blyton and P. Turnbull (eds) *Reassessing Human Resource Management*, London: Sage Publications.

Kenyon, T. A., Valway, D. M. D., Ihle, W. W., Onorato, I. M. and Castro, K. G. (1996) 'Transmission of multidrug-resistant tuberculosis during a long airplane flight', *The New England Journal of Medicine* 334(15).

Kinnie, N., Hutchinson, S. and Purcell, J. (2000) 'Fun and surveillance: the paradox of high commitment management in call centres', *International Journal of Human Resource Management* 11(5): 967–85.

Kochan, T. and Dyer, L. (1993) 'Managing transformational change: the role of HRM professionals', *International Journal of Human Resource Management* 4(3): 569–91.

Kogi, K., Ong, C. N. and Cabantog, C. (1989) 'Some social aspects of shift work in Asian developing countries', *International Journal of Industrial Ergonomics* 4: 151–9.

Koplin, K. (2000) Evidence to the House of Lords Select Committee on Science and Technology: Air Travel and Health, 13 June.

Labour Market Statistics (2001) *National Statistics*, June, London.

Labour Market Trends (1997) *Temporary Workers in Great Britain*, September.

Labour Research (1997) *Putting Safety Back Into Politics*, April: 19–20.

Labour Research Council (1999) *Indecent Exposure*, A joint report on noise at work commissioned by the Royal National Institute for Deaf People (RNID) and the Trades Union Congress (TUC), March.

Lalande, N. M. (1986) 'Is occupational noise exposure during pregnancy a risk factor of damage to the auditory system of the foetus?', *American Journal of Industrial Medicine* 10: 427–35.

Lankshear, G., Cook, P., Mason, D., Coates, S. and Button, G. (2001) 'Call centre employees' responses to electronic monitoring: some research findings', *Work, Employment and Society* 15(3): 595–605.

Laundry, B. R. and Lees, R. E. M. (1991) 'Industrial accident experience of one company on 8 and 12 hour shift systems', *Journal of Occupational Medicine* 33: 903–6.

Lebuser, H. J., Krasher, E. and Nubohm, E. (1995) 'Exposure of aircraft crews and frequent fliers to radiation', *Cockpit*, July: 14–19.

Lee, T. (1998) 'Assessment of safety culture at a nuclear reprocessing plant', *Work and Stress* 12(3): 217.

Legge, K. (1995) *Human Resource Management: Rhetoric and Reality*, London: Macmillan.

Leigh, J. P. (1982) 'Are unionised blue collar jobs more hazardous than non-unionised blue collar jobs?', *Journal of Labor Research* 3(3).

Lindgren, A. and Sederblad, P. (2000) 'Work organisation, control and qualifications in a travel agency and in a call centre', paper presented to the 18th Annual International Labour Process Conference, 25–7 April, University of Strathclyde.

Litwin, A. S. (2000) 'Trade Unions and Industrial Injury in Great Britain', Discussion Paper 468, Centre of Economic Performance, London School of Economics and Political Science.

London Health Education Authority (1997) *What People Think About Air Pollution, Their Health in General, and Asthma in Particular*, London: Health Education Unit.

Lowden, A., Kecklund, G., Axelsson, J. and Åkerstedt, T. (1999) 'Effects of 8 and 12 hours shifts on sleep and sleepiness and performance', Working Paper, National Institute of Working Life, Sweden, Health Hazards in the New Working Life Conference, 11–13 January, Stockholm.

Lowden, A. and Åkerstedt, T. (1999) 'Eastward long distance flights, sleep and wake patterns in air crews in connection with a two-day layover', *Journal of Sleep Research* 8(1): 15–24.

Lynge, E. (1996) 'Risk of breast cancer is also increased among Danish female airline cabin attendants', *British Medical Journal* 312(7025): 253.

Lyon, D. (1988) *The Information Society: Issues and Illusions*, Oxford: Polity Press.

Macdonald, C. L. and Sirianni, C. (1996) 'The service society and the changing experience of work', in C. L. Macdonald and C. Sirianni (eds) *Working in the Service Society*, Philadelphia: Temple University Press.

Mackay, C. J. and Cooper, C. L. (1987) 'Occupational stress and health: some current issues', in C. L. Cooper and I. T. Robertson (eds) *International Review of Industrial and Organizational Psychology*, Chichester: John Wiley and Sons.

Mcfarland, J. W., Hickman, C., Osterholm, M. T. and McDonald, K. L. (1994) 'Exposure to mycobacterium tuberculosis during air travel', *Lancet* 342: 112–13.

McLaughlin, C. and Rasmussen, E. (1998) '"Freedom of choice" and "flexibility" in the retail sector?', *International Journal of Manpower* 19(4).

Manufacturing, Science and Finance Union (MSF) (1997) *Work-Related Illness*, London: MSF.

Marginson, P., Armstrong, P., Edwards, P., Purcell, J. and Hubbard, N. (1993) *The Control of Industrial Relations in Large Companies: An Initial Analysis of the Second Company Level Industrial Relations Survey*, Warwick Papers in Industrial Relations, No. 45, University of Warwick.

Maruyama, S., Kohno, K. and Morimoto, K. (1995) 'A study of preventative medicine in relation to mental health among middle-management employees (part 2) – effects of long working hours on lifestyles, perceived stress and working-life satisfaction among white-collar middle-management employees', *Japanese Journal of Hygiene* 50: 849–60.

Maslach, C. (1982) *Burnout: The Cost of Caring*, Englewood Cliffs, NJ: Prentice Hall.

Maslach, C. and Jackson, S. E. (1985) 'The role of sex and family variable in burnout', in *Sex Roles* 12: 837–51.

Maslach, C. and Lieter, M. (1997) *The Truth About Burnout*, San Francisco: Jossey-Bass.

Mason, K. T. (1994) *Pregnancy and Flying Duties*, Aircrew Protection Division, United States Army, Aeromedical Research Laboratory, Fort Rucker, Alabama, Report No. 94, August 1994.

Mawson, A. R. (1998) 'Breast cancer in female flight attendants', *Lancet* 352.

Mayhew, C. (2002) 'Occupational violence in industrialised countries: types, incidence patterns and "at risk" groups of workers', in M. Gill, B. Fisher and V. Bowie (eds) *Violence at Work: Causes, Patterns and Prevention*, Devon: Willian Publishing.

Mayhew, C. and Quinlan, M. (1997a) 'Subcontracting and OHS in the residential building sector', *Industrial Relations Journal* 28(3): 192–205.

Mayhew, C. and Quinlan, M. (1997b) 'Trucking tragedies: why OHS outcomes are worse for subcontract workers in the road transport industry', Working Paper No. 114, School of Industrial Relations and Organisational Behaviour, University of New South Wales.

Mayhew, C. and Quinlan, M. (1999) 'The relationship between precarious employment and patterns of occupational violence: survey evidence from thirteen occupations', paper presented at the Health Hazards and Challenges in the New Working Life Conference, 11–13 January, Stockholm.

Mayhew, C., Quinlan, M. and Ferris, R. (1997) 'The effects of subcontracting/outsourcing in the Australian clothing industry: survey evidence from four Australian industries', *Safety Science* 25(1–3): 163–78.

Melton, C. E. (1982) 'Effects of long-term exposure to low levels of ozone and acid chlorides', *Aviation, Space and Environmental Medicine* 53: 105–11.

Mercer, A. and Brown, J. D. (1998) 'Venous thromboembolism associated with air travel', *Aviation, Space and Environmental Medicine* 69(2): 154–7.

Millward, N., Bryson, A. and Forth, J. (2000) *All Change at Work? British Employment Relations 1980–1998*, as portrayed by the Workplace Industrial Relations Survey series, London: Routledge.

Millward, N., Stevens, M., Smart, D. and Hawes, W. (1992) *Workplace Industrial Relations in Transition*, Dartmouth: Aldershot.

Moody, K. (1987) 'Go-it-alone mentality hurts airline unions in era of deregulation', *Labor Notes*, June: 8–9.

Moore, R. (1991) *The Price of Safety: The Market, Workers' Rights and the Law*, London: The Institute of Employment Rights.

Morris, J. A. and Feldman, D. C. (1996) 'The dimensions, antecedents and consequences of emotional labor', *Academy of Management Journal* 21: 989–1010.

Morris, J. A. and Feldman, D. C. (1997) 'Managing emotions in the workplace', *Journal of Managerial Issues* 9: 257–74.

Most, I. G. (1999) 'Psychosocial elements in the work environment of a large call centre operation', *Occupational Medicine* 14(1): 135–47.

Nagda, N. L., Rector, H. E., Zhidong, L. and Space, D. R. (2000) 'Aircraft cabin air quality: a critical review of past monitoring studies', in N. L. Nagad (ed.) *Air Quality and Comfort in Airliner Cabins, ASTM STP 1393*, West Conshohocken, PA: ASTM.

National Association of Citizens' Advice Bureaux (1997) *Flexibility Abused: a CAB Evidence Report on Employment Conditions in the Labour Market*, London: NCAB.

Nenot, J. C. (1998) 'Radiation accidents: lessons learnt for future radiological protection', *International Journal of Radiation Biology* 73(4): 435–42.

Nichols, T. (1986) 'Industrial injuries in British manufacturing in the 1980s', *The Sociological Review* 34(2): 290–306.

Nichols, T. (1997) *The Sociology of Industrial Injury*, London: Mansell.

Nichols, T. and Armstrong, P. (1973) *Safety or Profit: Industrial Accidents and the Conventional Wisdom*, Bristol: Falling Wall Press.

Norlén, U. and Andersson, K. (eds) *Bostadsbeståndets inneklimat*, Elib-rapport Nr.7, Gävle: Statens Institut för Byggnadsforskning.

Northwest Coalition for Alternatives to Pesticides (NCAP) *Report on Airline Sprays* (1999) http://www.efn.org/~ncap/AirlineSpray.pdf

Nuclear Energy Agency (NEA) (1999) *Identification and Assessment of Organisational Factors Related to the Safety of NPPs*, NEA/CSNI/R99, Issy-les-Moulinex, France, September.

Nuclear Energy Agency (NEA) (2000) *Assuring Nuclear Safety Competence into the 21st Century*, Workshop Proceedings, Budapest, Hungary, 12–14 October 1999, Paris: OECD Publications.

Nuclear Installations Inspectorate (NII) (2001) 'A public opinion survey', in *Investing in Trust: Nuclear Regulators and the Public*, Workshop Proceedings, Paris, France, 29 November–1 December 2000, Paris: OECD Publications.

Observer, 19 May 1996.

Observer (2001), 'Airlines to issue DVT alert', 5 August: 1.

Observer (2001) 'Putting a price on justice', 5 August: 16.

Observer (2001) 'Toxic fumes in aircraft spur inquiry', 9 September: 9.

Observer (2002) 'BA hauls in flyers to check DVT risk', 14 April: 13.

Odaka, K. and Sawai, M. (1999) *Small Firms, Large Concerns: The Development of Small Business in Comparative Perspective*, Oxford: Oxford University Press.

Office for National Statistics (2000) *Employment Statistics*, December, London.

Oldaker, G. B. and Conrad, R. C. (1987) 'Estimation of effects of environmental tobacco smoke on air quality within passenger cabins of commercial aircraft', *Environmental Science Technology*, 1987 (21): 994–9.

Oliver, N. and Wilkinson, B. (1992) *The Japanisation of British Industry*, Oxford: Blackwell.

Ono, Y., Watanabe, S., Kaneko, S., Matsumoto, K. and Miyako, M. (1991) 'Working hours and fatigue of Japanese flight attendants', *Journal of Human Ergology* 20: 155–64.

Organisation for Economic Co-operation and Development (OECD) (1986) 'Occupational accidents in OECD countries', *Employment Outlook*, July, Paris: OECD.

Organisation for Economic Co-operation and Development (OECD) (1989) *Labour Market Flexibility: Trends in Enterprises*, Paris: OECD.

Organisation for Economic Co-operation and Development (OECD) (1997) *The OECD Jobs Study: Unemployment in the OECD Area, 1950–1995*, Paris: OECD.

Organisation for Economic Co-operation and Development (OECD) (2001) *Investing in Trust: Nuclear Regulators and the Public*, Workshop Proceedings, Paris, France, 29 November–1 December 2000, Paris: OECD.

Osborne, J. and Zairi, M. (1997) *Total Quality Management and the Management of Health and Safety*, HSE Contract Report 153, London: HSE.

Paules, G. F. (1996) 'Resisting the symbolism of service among waitresses', in C. L. Macdonald and C. Sirianni (eds) *Working in the Service Society*, Philadelphia: Temple University Press.

Pease, K. (1985) 'Obscene telephone calls to women in England and Wales', *Howard Journal of Criminal Justice* 24: 275–81.

Peccei, R. and Rosenthal, P. (1997) 'The antecedents of employee commitment to customer service: evidence from a UK service context', *International Journal of Human Resource Management* 8(1): 66–86.

Pfausler, B., Vollert, H., Bosch, S. and Schmutzhard, E. (1996) 'Cerebral venous thrombosis – a new diagnosis in travel medicine?', *Journal of Travel Medicine* 3: 165–76.

Pfeffer, J. (1997) *New Directions in Organisation Theory*, Oxford: Oxford University Press.

Pidgeon, N. (1998) 'Safety culture: key theoretical issues', *Work and Stress* 12(3): 202–16.

Pidgeon, N. and O'Leary, M. (1994) 'Organizational safety culture: implications for aviation practice', in N. A. Johnstone, N. McDonald and R. Fuller (eds) *Aviation Psychology in Practice*, Aldershot: Avebury Technical.

Pizzino, A. (1994) *Report on CUPE's (Canadian Union of Public Sector Employees) National Health and Safety Survey of Aggression Against Staff*, Ottawa: CUPE.

Pukkala, E., Auvinen, A. and Wahlberg, G. (1995) 'Incidence of cancer among Finnish airline cabin attendants, 1967–92', *British Medical Journal* 311, September: 649–51.

Quinlan, M. (1999) 'The implications of labour market restructuring in industrialized societies for occupational health and safety', *Economic and Industrial Democracy* 20(3): 427–60.

Quinlan, M. (2002) *Report of Inquiry into Safety in the Long Haul Trucking Industry: Executive Summary*, Industrial Relations Research Centre, University of New South Wales, Australia.

Rafnsson, V., Hrafnkelsson, J. and Tulinius, H. (2000) 'Skin cancers in airline pilots', *Occupational Health and Environmental Medicine* 57: 175–9.

Rafnsson, V., Tulinius, H., Jonasson, J. G. and Hrafnkelsson, J. (2001) 'Risk of breast cancer in female flight attendants: a population-based study', *Cancer Causes and Control* 12: 95–101.

Railway Industry Advisory Committee (RIAC) (1999) *Final Report of Conference on Violence to Staff on Railways*, 8 July, RIAC/HSC.

Ramsay, R. E. (1995) 'Involvement, empowerment and commitment', in B. Towers (ed.) *The Handbook of Human Resource Management*, Oxford: Blackwell.

Raw, G. (1992) *Sick Building Syndrome: A Review of the Evidence on Causes and Solutions*, HSE Contract Research Report, No. 42/1992, London: HSE.

Reason, J. (1990) 'The contribution of latent human failures to the breakdown of complex systems', in D. E. Broadbent, J. Reason and A. Baddeley (eds) *Human Factors in Hazardous Situations*, Oxford: Clarendon Press.

Reilly, B., Paci, P. and Holl, P. (1995) 'Unions, safety and workplace injuries', *British Journal of Industrial Relations* 33(2): 276–88.

Ribier, B., Zizka, V., Cysique, J., Danalien, Y., Glaudon, G. and Ramialison, C. (1997) 'Venous thromboembolic events following air travel', *Revue de Médecine Interne* 18(8): 601–4.

Richardson, R. and Belt, V. (2001) 'Saved by the bell? Call centres and economic development in less favoured regions', *Economic and Industrial Democracy* 22: 67–8.

Richardson, R., Belt, V. and Marshall, J. N. (2000) 'Taking calls to Newcastle: the regional implications of the growth in call centres', *Regional Studies* 34(4): 357–69.

Rijpma, J. A. (1996) 'Complexity, tight coupling and reliability: connecting normal accidents theory with high reliability theory', *Journal of Contingencies and Crisis Management* 4: 15–24.

Riley, B. (1999) 'Report highlights risk of pesticides used on aircraft', *Pesticide News* 43: 16.

Risen, J. (1987) 'Productivity push: peril on the job', *Los Angeles Times*, 2 March: Part 1(1): 12.

Ritzer, G. (1993) *The McDonaldization of Society*, Thousand Oaks, CA: Sage Publications Inc.

Robens Report (1972) *Safety and Health at Work*, Cmnd. 5034, London: HMSO.

Robertson, G. (1989) *Testimony Before Sub-Committee on Aviation*, US House Committee on Public Works and Transport.

Robinson, J. and Nelson, W. C. (1995) *National Human Activity Pattern Survey Data Base*, United States Environmental Protection Agency, Research Triangle Park, NC.

Romano, E., Ferrucci, L., Nicolai, R., Derme, V. and De Stefano, G. F. (1997) 'Increase of chromosomal aberrations induced by ionising radiation in peripheral blood lymphocytes of civil aviation pilots and crew members', *Mutation Research – Fundamental and Molecular Mechanisms of Mutagenesis* 377(1): 89–93.

Roncoroni, S. (2000) 'Call centres in the new millennium', Call Centre Association 6th Annual Convention, 14 November, Glasgow.

Rosa, R. R. (1995) 'Extended workshifts and excessive fatigue', *Journal of Sleep Research* 4(2): 51–6.

Rutter, D. R. and Fielding, P. J. (1988) 'Sources of occupational stress: an examination of British police officers', *Work and Stress* 2: 291–9.

Sagan, S. D. (1993) *The Limits of Safety: Organizations, Accidents and Nuclear Weapons*, Princeton: Princeton University Press.

Saliminen, S., Saari, J., Saarela, K. and Rasanen, T. (1993) 'Organisational factors influencing serious occupational accidents', *Scandinavian Journal of Work, Environment and Health* 19: 352–7.

Saxton, M. J., Phillips, J. S. and Blakeney, R. N. (1991) 'Antecedents and consequences of emotional exhaustion in the airline reservations service sector', *Human Relations* 44(6): 581–95.

Schaufeli, W. B. and Buunk, B. P. (1996) 'Professional burnout', in M. J. Schabracq, J. A. M. Winnubust and C. L. Cooper (eds) *Handbook of Work and Health Psychology*, Chichester: John Wiley and Sons.

Schaufeli, W. B. and Enzmann, D. (1998) *The Burnout Companion to Study and Practice: A Critical Analysis*, London: Taylor and Francis.

Scheid, W., Weber, J., Traut, H. and Gabriel, H. W. (1993) Institut für Strahlenbiologie der Universität Münster, *Naturwissenschaften* 80: 530–8.

Schnieder and Bowen (1993) 'The service organisation: human resource management is crucial', *Organisational Dynamics* 21(4): 39–52.

Schrewe, U. J. (2000) 'Global measurements of the radiation exposure of civil air crew from 1997', *Radiological Protection Dosimetry* 91: 347–64.

Schuler, R. (1980) 'Definition and Conceptualisation of Stress in Organisations', *Organisational Behaviour and Human Decision Processes* 25: 184–215.

Scurr, J. H., Machin, S. J., Bailey-King, S., Mackie, I. J., McDonald, S. and Coleridge Smith, P. D. (2001) 'Frequency and prevention of symptomless deep-vein thrombosis in long-haul flights: a randomised trial', *Lancet* 357: 1485–9.

Sczesny, S. and Stahlberg, D. (1999) *Sexuelle Belaestigung am Telefon: Eine sozialpsychologische Analyse*, Frankfurt: Lang.

Sczesny, S. and Stahlberg, D. (2000) 'Sexual harassment over the telephone: occupational risk at call centres', *Work and Stress* 14(2): 121–36.

Sewell, G. and Wilkinson, B. (1992) 'Empowerment or emasculation? Shopfloor surveillance in a total quality organisation', in P. Blyton and P. Turnbull (eds) (1995) *Reassessing Human Resource Management*, London: Sage.

Sheffied, C. J. (1989) 'The invisible intruder: women's experiences of obscene telephone calls', *Gender and Society* 3: 483–8.

Sisson, K. (1994) *Personnel Management*, 2nd edition, Oxford: Blackwell (ref. to Bach 1994).

Smith, A. J. (1996) 'Cabin air quality: what is the problem? What is being done or what can be done about it? Who can do it and how?', *Journal of Air Law and Commerce* 61(3), February.

Smith, M. J. and Cohen, H. H. and Cohen, A. (1978) 'Characteristics of a successful safety program', *Journal of Safety Research* 19: 5–15.

Smith, M. D. and Morra, N. N. (1994) 'Obscene and threatening phone calls to women: data from a Canadian national survey', *Gender and Society* 8: 584–96.

Society of Radiographers (1991) *Preventing Darkroom Disease*, London.

Sparks, K., Cooper, C., Fried, Y. and Shirom, A. (1997) 'The effects of hours of work on health: a meta-analytic review', *Journal of Occupational and Organizational Psychology* 70: 391–408.

Stanworth, C. (2000) 'Women and work in the information age', *Gender, Work and Organization* 7(1): 20–32.

Stenberg, B. and Wall, S. (1995) 'Why do women report "sick building syndrome" more often that men?', *Social Sciences Medicine* 40(4): 491–502.

Stonier, T. (1983) *The Wealth of Information*, London: Thames-Methuen.

Storey, J. (1987) 'Developments in the management of human resources: an interim report', *Warwick Papers in Industrial Relations* 17, IRRU, School of Industrial and Business Studies, University of Warwick.

Storey, J. (1995) *Human Resource Management: A Critical Text*, London: Routledge.

Suvanto, S., Partinen, M., Harma, M., Ilmarinen, J. (1990) 'Flight attendants desynchronosis after rapid time zone changes', *Aviation, Space and Environmental Medicine* 61: 543–7.

Sznelwar, L. I., Mascia, F. L. and Zilbovicius, M. (1999) 'Ergonomics and work organization: the relationship between Tayloristic design and workers' health in banks and credit card companies', *International Journal of Occupational Safety and Ergonomics* 5(2): 291–301.

Szubert, Z., Sobala, W. and Zycińska, A. (1997) 'The effect of system restructuring on absenteeism due to sickness in the workplace', *Medycyna Pracy* 48(5): 543–51.

Taylor, F. W. (1947) 'The principles of scientific management', in F. W. Taylor (ed.) *Scientific Management*, New York: Harper.

Taylor, P. and Bain, P. (1999) 'An assembly line in the head: work and employee relations in the call centre', *Industrial Relations Journal* 30(2): 101–17.

Taylor, P. and Bain, P. (2001) 'Trade unions, workers' rights and the frontier of control in UK call centres', *Economic and Industrial Democracy* 22: 39–66.

Taylor, S. (1998) 'Emotional labour and the new workplace', in P. Thompson and C. Warhurst (eds) *Workplaces of the Future*, London: Macmillan.

Teenan, R. P. and Mackay, A. J. (1992) 'Peripheral arterial thrombosis related to commercial airline flights: another manifestation of the economy class syndrome', *British Journal of Clinical Practice* 46(3): 165–6.

Thiebault, C. (1997) 'Cabin air quality', *Aviation, Space and Environmental Medicine* 68(1): 80–2.

Thomas, G. (2000) Evidence to the House of Lords Select Committee on Science and Technology: Air Travel and Health, 20 June 2000.

Thompson, P. and Warhurst, C. (eds) (1998) *Workplaces of the Future*, London: Macmillan.

Thörn, A. (1998) 'The sick building syndrome: a diagnostic dilemma', *Social Sciences Medicine* 47(9): 1307–12.

Toffler, A. (1980) *The Third Wave*, London: Pan Books.

Towers, B. (2000) Notes from Professor Towers' speech at the ACAS Fringe Meeting 'The end of industrial relations as we know it?', TUC Congress, 11 September, http/www.acas.org.uk

Trades Union Congress (TUC) (1995) 'Absence', London: TUC.

Trades Union Congress (TUC) (1999a) 'Focus on balloting and industrial action', *Trade Union Trends Survey*: 3.

Trades Union Congress (TUC) (1999b) 'Violent Times', January, London (chaps 4, 5).

Trades Union Congress (TUC) (1999c) 'Going to work can damage your hearing', Press Release, 16 March.

Trades Union Congress (TUC) (2000) 'Focus on balloting and industrial action', *Trade Union Trends Survey*: 2.

Trades Union Congress (2001a) 'Focus on balloting and industrial action', *Trade Union Trends Survey*: 4.

Trades Union Congress (2001b) 'Calls for change', Second TUC report of calls to the 'It's Your Call' hotline, April, London: TUC.

Transport and General Workers' Union (TGWU) (1998) Information provided during consultation over cabin-crew survey.

Truss, C., Gratton, L., Hope-Hailey, V., McGovern, P. and Stiles, P. (1997) 'Soft and hard models of human resource management: a reappraisal', *Journal of Management Studies* 34(1): 53–73.

Turiel, I., Hollowell, C. D., Biksch, R. R., Rudy, J. V., Young, R. A. and Coye, M. J. (1983) 'The effects of reduced ventilation on indoor air quality in an office building', *Atmospheric Environment* 17: 51–64.

Turner, G. and Myerson, J. (1998) *New Workspace, New Culture: Office Design as a Catalyst for Change*, Aldershot: Gower.

Turner, B. A., Pidgeon, N. F., Blockley, D. I. and Toft, B. (1989) 'Safety culture: its importance in future risk management', Position paper for the Second World Bank Workshop on Safety Culture and Risk Management, November, Karlstad, Sweden.

UNISON (1998) *Control or Management*, London.

Vahtera, J., Kivimaki, M. and Pentti, J. (1998) 'Effects of organisational downsizing on health of employees', *Lancet* 350: 1124–8.

Van Maanen, J. and Kunda, G. (1989) 'Real feelings: emotional exhaustion and organisational culture', *Research in Organisational Behaviour* 11: 43–103.

Vasak, V. (1986) *The Airliner Cabin Environment. Air Quality and Safety*, National Research Council Study, Washington DC.

Vassie, L. (1998) 'A proactive team-based approach to continuous improvement in health and safety management', *Employee Relations* 20(6): 577–93.

Vaughan, T. L., Daling, J. R. and Starzyk, P. M. (1984) 'Fetal death and maternal occupation: an analysis of birth records in the state of Washington', *Journal of Occupational Medicine* 26: 676–8.

Viscusi, W. Kip (1979) *Employment Hazards: An Investigation of Market Performance*, Cambridge, MA: Harvard University Press.

Waddingon, J. and Whitson, C. (1996) 'Empowerment versus intensification: union perspectives of change at the workplace', in P. Ackers, C. Smith and P. Smith (eds) *The New Workplace of Trade Unionism*, London: Routledge.

Wagner, D. R. (1996) 'Disorders of the circadian sleep–wake cycle', *Neurol. Clin.* 14(3): 651–70.

Wallace, C. M., Eagleson, G. and Waldersee, R. (2000) 'The sacrificial HR strategy in call centres', *International Journal of Service Industry Management* 11(2): 174–84.

Wallace, L. A. (1997) 'Sick building syndrome', in E. J. Bardana and A. Montanaro (eds) *Indoor Air Pollution and Health*, New York: Marcel Dekker.

Walters, D. (1996) 'Trade unions and the effectiveness of worker representation in health and safety in Europe', *Employee Relations* 18(6): 48–66.

Walton, R. E. (1985) 'Toward a strategy of eliciting employee commitment based on policies of mutuality', in R. W. Walton and P. R. Lawrence (eds) *Human Resource Management, Trends and Challenges*, Boston: Harvard Business School Press.

Warhurst, C. and Thompson, P. (1998) 'Hands, hearts and minds: changing work and workers at the end of the century', in P. Thompson and C. Warhurst (eds) *Workplaces of the Future*, London: Macmillan.

Warhurst, W. (1995) 'Converging on HRM? Change and continuity in European airlines industrial relations', *European Journal of Industrial Relations* 1(2): 266.

Waters, M., Bloom, T. F. and Grajewski, B. (2000) 'The NIOSH/FAA working women's health study: evaluation of the cosmic-radiation exposures of flight attendants', *Health Physics* 79: 553–9.

Wharton, A. S. (1993) 'The affective consequences of service work', *Work and Occupations* 20: 205–32.

Wharton, A. S. (1996) 'Service with a smile: understanding the consequence of emotional labor', in C. L. Macdonald and C. Sirianni (eds) *Working in the Service Society*, Philadelphia: Temple University Press.

Wiatrowski, W. (1994) 'Small businesses and their employees', *Monthly Labour Review* 117(10): 29–35.

Wilpert, B. and Itoigawa, N. (2002) *Safety Culture in Nuclear Power Operations*, London: Taylor and Francis.

Witkowski, C. (1999) 'Organophospate hazards', Working Paper presented at the International Transport Workers' Federation Cabin Crew Health and Safety Conference, 20–22 April, Amsterdam.

Witmer, E., Minister for Labour (1997) 'Review of the Occupational Health and Safety Act', Discussion Paper, February, Ontario.

Wokutch (1990) *Co-operation and Conflict in Occupational Safety and Health: A Multinational Study of the Automotive Industry*, New York: Praeger.

Wood, S. and de Menezes, L. (1998) 'High commitment management in the UK: evidence from the Workplace Industrial Relations Survey and Employers' Manpower and Skills Practices Survey', *Human Relations* 51(4): 485–515.

Woolfson, C. and Beck, M. (1995) 'Deregulation: the contemporary politics of health and safety', in E. McCogan (ed.) *The Future of Labour Law*, London: Mansell, 171–205.

Woolfson, C. and Beck, M. (1998) *From Self-Regulation to Deregulation: The Politics of Health and Safety in Britain*, University of Glasgow/University of St Andrews.

World Health Organisation (WHO) (1999) *Tuberculosis and Air Travel: Guidelines for Prevention and Control*, Geneva: WHO.

Worrall, J. D. and Butler, R. J. (1983) 'Health conditions and job hazards: union and non-union jobs', *Journal of Labor Research* 4(4).

Zapf, D., Vogt, C., Seifert, C., Mertini, H. and Isic, A. (1999) 'Emotion work as a source of stress: the concept and development of an instrument', *European Journal of Work and Organizational Psychology* 8(3): 371–400.

Zuboff, S. (1988) *In the Age of the Smart Machine: The Future of Work and Power*, New York: Basic Books.

Author index

Subject index

Printed in the United Kingdom
by Lightning Source UK Ltd.
118738UK00006B/253